AF803520

HOUSING, SPACE AND QUALITY OF LIFE

Housing, Space and Quality of Life

Edited by

RICARDO GARCÍA MIRA
University of Corunna, Spain

DAVID L. UZZELL
University of Surrey, UK

J. EULOGIO REAL
University of Santiago de Compostela, Spain

JOSÉ ROMAY
University of Corunna, Spain

Routledge
Taylor & Francis Group
LONDON AND NEW YORK

First published 2005 by Ashgate Publishing

Reissued 2018 by Routledge
2 Park Square, Milton Park, Abingdon, Oxon OX14 4RN
711 Third Avenue, New York, NY 10017, USA

Routledge is an imprint of the Taylor & Francis Group, an informa business

© Ricardo García-Mira, David L. Uzzell, J. Eulogio Real and José Romay 2005

Ricardo García-Mira, David L. Uzzell, J. Eulogio Real and José Romay have asserted their right under the Copyright, Designs and Patents Act, 1988, to be identified as the editors of this work.

All rights reserved. No part of this book may be reprinted or reproduced or utilised in any form or by any electronic, mechanical, or other means, now known or hereafter invented, including photocopying and recording, or in any information storage or retrieval system, without permission in writing from the publishers.

A Library of Congress record exists under LC control number: 2004027709

Notice:
Product or corporate names may be trademarks or registered trademarks, and are used only for identification and explanation without intent to infringe.

Publisher's Note
The publisher has gone to great lengths to ensure the quality of this reprint but points out that some imperfections in the original copies may be apparent.

Disclaimer
The publisher has made every effort to trace copyright holders and welcomes correspondence from those they have been unable to contact.

ISBN 13: 978-0-815-38952-1 (hbk)
ISBN 13: 978-1-351-15636-3 (ebk)

Contents

List of Contributors

Leanne Abbott, The Scott Sutherland School, The Robert Gordon University, Aberdeen, Scotland, AB10 7QB, United Kingdom. l.c.abbott@rgu.ac.uk

Suat Apak, Faculty of Architecture, Istanbul Technical University, Taksim 80191, Istanbul, Turkey. aunlu@itu.edu.tr

Stelamaris R. Bertolli, Department of Architecture and Construction, School of Civil Engineering, State University of Campinas, UNICAMP,CP 6021 – 13084-971 Campinas, SP. Brasil. rolla@fec.unicamp.br

Francisco Borges Filho, Department of Architecture and Construction, School of Civil Engineering, State University of Campinas, UNICAMP,CP 6021, 13084-971 Campinas. SP. Brasil

Rodrigo Cantarero, Department of Community and Regional Planning, University of Nebraska, 302 Architecture Hall, Lincoln NE 68588-0105, USA. rcantarero@unl.edu

Anthony Craig, The Scott Sutherland School, The Robert Gordon University, Aberdeen, Scotland, AB10 7QB, United Kingdom. a.craig@rgu.ac.uk

Martin Edge, The Scott Sutherland School, The Robert Gordon University, Aberdeen, Scotland, AB10 7QB, United Kingdom. m.edge@rgu.ac.uk

Édison Fávero, Department of Architecture and Construction – School of Civil Engineering, State University of Campinas, UNICAMP, CP 6021, 13084-971 Campinas, SP. favero@fec.unicamp.br

Ángel Fernández González, Social Psychology Department, Faculty of Educational Sciences, University of Vigo, Spain, Campus of As Lagoas, 32004, Ourense, Spain. afdez@uvigo.es

Ricardo García-Mira, Department of Psychology, Faculty of Educational Sciences, University of Corunna, Campus of Elviña, 15071, A Coruña, Spain. fargmira@udc.es

Myriam Goluboff, Faculty of Architecture, Campus of A Zapateira, 15071, A Coruña, Spain. mygolu@udc.es

Heather Keele, College of Architecture, University of Nebraska, 239 Architecture Hall, Lincoln, NE 68588-0107, USA. heatherdawn@juno.com

Doris C.C.K. Kowaltowski, Department of Architecture and Construction, School of Civil Engineering, State University of Campinas, UNICAMP, CP 6021, 13084-971 Campinas, SP. doris@fec.unicamp.br

Lucila C. Labaki, Department of Architecture and Construction School of Civil Engineering, State University of Campinas, UNICAMP, CP 6021 - 13084-971 Campinas, SP. Brasil. lucila@fec.unicamp.br

Richard Laing, The Scott Sutherland School, The Robert Gordon University, Aberdeen, Scotland, AB10 7QB, United Kingdom. r.laing@rgu.ac.uk

Steven Larrick, Community Development Coordinator, University of Nebraska, Lincoln, NE 68588-0107, USA. slarrick1@unl.edu

Mariann Märtsin, Department of Psychology, Tallin Pedagogical University, Narva maantee 25, EE-10120 Tallin, Estonia. tniit@tpu.ee

Silvia A. Mikami G. Pina, Department of Architecture and Construction School of Civil Engineering, State University of Campinas, UNICAMP, CP 6021, 13084-971 Campinas, SP. Brasil. smikami@fec.unicamp.br

Toomas Niit, Department of Psychology, Tallin Pedagogical University, Narva maantee 25, EE – 10120 Tallin, Estonia. tniit@tpu.ee

Harmen Oppewal, Department of Marketing, Faculty of Business and Economics, Monash University, PO Box 197, Caulfield East, VIC 3145, Australia. Harmen.Oppewal@BusEco.monash.edu.au

Ritsuko Ozaki, The Business School, Imperial College London, South Kensington Campus, London SW7 2AZ, United Kingdom. r.ozaki@imperial.ac.uk

Ahsen Özsoy, Faculty of Architecture, Istanbul Technical University, Taskısla, Taksim, 80191, Istanbul, Turkey. ozsoya@itu.edu.tr

Yaniv Poria, Department of Hotel and Tourism Management, Ben Gurion, University of the Negev, Beer-Sheva, Israel. Yporia@bgumail.bgu.ac.il

James J. Potter, College of Architecture, University of Nebraska, 239 Architecture Hall, Lincoln, NE 68588-0107, USA. jpotter2@unl.edu

Gulçin Pulat Gökmen, Faculty of Architecture, Istanbul Technical University, Taskısla, Taksim, 80191, Istanbul, Turkey. ggokmen@itu.edu.tr

Blanca E. Ramirez, Nebraska Equal Opportunities Commission, 301 Centennial Mall South, 5th P.O. Box 94934 Lincoln, NE 68509-4934, USA. bramirez@neoc.state.ne.us

Neil Ravenscroft, Chelsea School Research Centre, University of Brighton, Trevin Towers Annex, Eastbourne, BN20 7SP, United Kingdom. N.Ravenscroft@bton.ac.uk

J. Eulogio Real, Department of Methodos and Research Techniques, Faculty of Psychology, University of Santiago de Compostela, Campus Sur, 15706 Santiago de Compostela, Spain. mtredeus@usc.es

José Romay, Department of Psychology, Faculty of Educational Sciences, University of Coruña, Campus de Elviña, 15071 A Coruña, Spain. romay@udc.es

Ombretta Romice, Department of Architecture, University of Strathclyde, 131 Rottenrow, Glasgow G4 ONG, United Kingdom. r.romice@strath.ac.uk

Regina C. Ruschel, Department of Architecture and Construction School of Civil Engineering, State University of Campinas, UNICAMP, CP 6021, 13084-971 Campinas, SP.

Gerda Speller, Department of Psychology, University of Surrey, Guildford, Surrey, GU2 7XH, United Kingdom. gerdaspellercons@btconnect.com

William J. Thompson, School of the Built Environment, University of Ulster, Newtownabbey, Co. Antrim, BT37 0QB. wj.thompson@ulster.ac.uk

Gokhan Ulken, Faculty of Architecture, Istanbul Technical University, Taksim 80191, Istanbul, Turkey.

Alper Unlu, Faculty of Architecture, Istanbul Technical University, Taksim 80191, Istanbul, Turkey. aunlu@itu.edu.tr

David L. Uzzell, Department of Psychology, University of Surrey, Guildford, Surrey, GU2 5XH, United Kingdom. d.uzzell@surrey.ac.uk

X. Winston Yan, Department of Architecture, University of Nebraska, Lincoln, NE 68588-0107, USA.

Acknowledgements

The editors express their gratitude to the following institutions and organizations for their cooperation in and support of this project:

Ayuntamiento de La Coruña
Concello de A Coruña

MINISTERIO DE MEDIO AMBIENTE

XUNTA DE GALICIA

FUNDACION CAIXAGALICIA

Coca-Cola BEGANO, S.A.

REPSOL YPF

Chapter 1

Housing, Space and Quality of Life: Introduction

Ricardo García-Mira, David L. Uzzell, J. Eulogio Real and José Romay

As a physical setting, the residential environment is critical for human well-being. We spend most of our working time in buildings, and most of our leisure time at home or close by in our neighbourhood. This fact alone justifies the need to study the role that housing and space play in the quality of life of individuals and communities.

The concept of quality of life is complex, because it includes a multitude of contributory facets such as housing, education, work and environment, as Blanco and Chacón (1985) point out. One can identify at least three different approaches to the study of quality of life in the context of housing and the environment. First, quality of life studies have focused on subjective well-being or life satisfaction (Donovan and Halpern, 2002). This research concentrates on asking people if they are satisfied with their lives in general, although it can be extended to examine individuals' longer term life goals and aspirations, as well as measuring people's self-reported psychological health and mental state. Second, quality of life has typically been understood by governments to be synonymous with standard of living (Jackson, 2002). Consequently, if a government strives to improve the nation's standard of living, it can be said to be improving the nation's quality of life. This is, of course, debatable especially in the context of the third interpretation of quality of life which has been to link the concept directly to sustainable development such that the two terms are used almost interchangeably. Partly this follows a philosophical argument that unless we engage in more sustainable practices the quality of life for the population will deteriorate; one suspects however, that the term quality of life has been used because it is more accessible and meaningful than the phrase sustainable development. Each of these three positions is reflected in the papers in this book.

One of the more important factors contributing to quality of life is housing, because it often serves to define the life space of a person. It has, however, been appreciated that housing, space and quality of life necessarily require a multi-disciplinary approach and this is reflected by research in recent years investigating, for example, planning participation (Horelli, 2004), the use of objective and subjective measures in modelling of residential quality and the design of indicators (Bonaiuto, Fornara and Bonnes, 2003; Marans, 2000, 2004), the study of space and its utilisation in the domestic context (Kellet, 2004), the study of space with regard to cultural diversity (Turgut and Kellet, 2001), the use of new technologies in communication and information (Craig and Edge, 2004), or data analysis and research methodology in general (Lawrence 1987, 2004; Hurol, Y., Urban Vestbro and Wilkinson, 2004).

The chapters in this book present the work of thirty-five researchers concerned with the relationship between housing, space and quality of life, in the context of the physical, psychological and social aspects of urban life.

Cities are not static phenomena; they are constantly changing. The rise of 'edge city' has seen the thrust of urban development in recent years occurring on the periphery of our cities, where retail and service facilities are leading to changes in architectural styles, economic patterns and social behaviour (Garreau, 1991; Rowe, 1991; Uzzell, 1995). Such significant changes lead to changes in the perception and interpretation of urban structure and spaces, which inevitably has an impact on people-environment relationships in general and how residents interact with the city in particular. This is addressed by García Mira and Goluboff in their analysis of the interpretation of urban space from the perspectives of two different types of users: pedestrians and automobile passengers. They demonstrate how the use of the car and other means of transport in cities will affect the knowledge of the urban environment and the perception and use of space.

The home is clearly a crucial aspect of the study of housing and space. The concept of home represents the essence of the housing experience. It may express the personality, culture and lifestyle of the homeowners. This, in turn, will depend upon whether the individual lives alone or shares (i.e., with family or friends), whether they are the owner or a renter, and their cultural background. The same physical space can be utilized in a different manner by different users. It is impossible to overestimate the importance of home as it is one of the principal sources of retreat, relaxation and social interaction in our leisure time. The chapters by Ozsoy and Pulat, and Ozaki address facets of the home in two different cultures, and its relationship to evaluations of quality of life and lifestyle.

But a home and the feeling of home is only the final step in the process of finding a place to live. Prior to that sense of home is the decision where to live and in what kind of house or apartment. This will be contingent upon objective economic constraints as well as subjective social and environmental preferences. The chapter by Oppewal *et al.* examines these issues in the case of student preferences for university accommodation. A further chapter on residential preferences by Craig *et al.* focuses more particularly on preferences for architectural properties and assesses preferences for different cladding materials on houses.

Choice in housing is the privilege of the wealthy. Many people do not have the luxury of choice – for them it is enough simply to find a place to live and to have a roof over their heads. This is, of course, not an uncommon experience in Third World countries. The chapter by Mikami *et al.* evaluates a technical and assistance programme which has the objective of helping Brazilian people who wish to move to a new location and self-build their houses in a safer and healthier way, while at the same time protecting the environment.

Houses and homes cannot be assessed simply in their own terms, as if separate and unrelated to their surroundings. The context in which houses are situated can be important not only for how they are appraised architecturally, but also the effect that location and setting can have on the evaluation of our quality of life. We may like our house, but if we are not happy with the surroundings this will have a negative effect on our overall evaluation of our housing and our quality of life. The chapter by Apak *et al.* evaluates a large residential area in Istanbul, and the problems related to security as a consequence of the spatial (e.g., configuration, inner or outer location) and social (e.g., use, isolation) characteristics of the urban area. Romice, drawing on the results of a European Union funded research project (NEHOM), also discusses the issue of quality of life from a neighbourhood perspective, arguing that integrated and coordinated actions are the most appropriate way to achieve successful urban management and the sustaianble regeneration of disadvantaged neighbourhoods. She concludes with a discussion of how current practice and research in interdisciplinary training and education of design and related disciplines in the UK can and should impact upon this kind of work.

Migration movements can affect significantly the perception of residential areas, and the consequential demographic changes may be evaluated differently by newcomers and long-time residents. This is particularly true for small communities, where the absorption of a new population is more problematic and the resources scarce. This is the subject of the chapter by Potter *et al.* who examine a small community in Nebraska.

The chapter by Thompson approaches the issue from a more global and interdisciplinary perspective, paying attention to the links between the buildings and their mental connections with individuals and groups. The development of mental representations of the spatial environment is also addressed in the chapter by Fernández. He argues the case for age as one of the relevant predictors of spatial knowledge and demonstrates how the interaction between children and their environment affects their acquisition and future development of environmental knowledge.

Finally, the chapter by Märtsin and Niit investigates the differences in the use of the home as a reflection of the socio-cultural context into which it is integrated with the norms and customs adopted by each culture. Home is analysed as a regulator of people's openness/closedness, and their management of the boundary between the self and non-self. The authors review the different theoretical concepts that have been employed for defining home as a territorial space comprising different functional units (e.g., public, family and private rooms).

Despite the efforts made over the last fifty years to develop a substantial and coherent theoretical framework to understand and analyse the design and construction of housing, the reality is that research efforts became fragmented as they were subject to the different disciplines involved in housing studies (i.e., architecture, psychology, urban planning). Furthermore, research has had difficulties connecting with the professional field of architecture, as has been pointed out by Symes (1984). One can identify at least two reasons why the social scientific study of housing has not been well integrated into professional architecture. First, there are the inevitable difficulties when two or more disciplines come together; not only will each have its own concepts, theories and methods; different disciplines have different languages and these can be significant barriers to communication. But even more important, each will hold particular assumptions about people-environment relationships. For example, one criticism that has been levelled at architects is that there has been a tendency to assume a deterministic perspective, i.e., architects tend to look to psychologists and social scientists for advice on how physical design can directly influence behaviour; social science deals with probabilities, not certainties. Second, there may be a conflict between the knowledge generation goals of science and the objective of application required by the design professions; science defines the problem narrowly in order to secure knowledge generation such that this makes the knowledge so context specific it may not be generalisable for practical application. Alternatively, the scientific search for general principles is at the expense of specific applicability (Gifford, 1998; Gärling and Hartig, 2000; Uzzell, 2000; Moser, 2000). Consequently, integrating into professional practice knowledge that is derived from user or residents' perceptions is not unproblematic.

It was also suggested by Proshansky, Ittelson and Rivlin (1970) that for the kind of interdisciplinarity that is being advocated here to work, it requires an interdisciplinary superstructure of evidence-based theoretical constructs. This can only emerge through cooperation and empirical endeavour. There is some evidence that this is now happening both in terms of practice and in the training and education of architects and psychologists (cf. Romice and Uzzell, forthcoming). This book offers another opportunity for a constructive and creative dialogue between the design disciplines and the social sciences in which theories and methodologies can be shared Drawing on presentations made at the 17th International Conference of the International Association for People-Environment Studies (2002, A Coruña, Spain) on the theme of *Culture, Quality of Life and Globalization - Problems and Challenges for the New Millennium*, all the chapters in this volume demonstrate the multi-disciplinary as well as interdisciplinary approaches that have been used to explore the contribution of housing and space to quality of life issues.

References

Blanco, A. and Chacón, F. (1985). La evaluación de la calidad de vida[The evaluation of the quality of life]. In J.F. Morales, A. Blanco, C. Huici y Fernández: *Psicología Social Aplicada [Applied Social Psychology]*. Bilbao: Desclée de Brouwer.

Bonaiuto, M., Fornara, F. Bonnes, M. (2003). Index of perceived residential environmental quality and neighbourhood attachement in the urban environment; a confirmatory study on the city of Rome. *Landscape and Urban Planning*, 65, 41-52.

Craig, A. and Edge, M. (2004). Internet-Based Methodologies in Housing Research: An Iterative Study Using Quantitative, Financial Measures to Gauge Housing Choices. In Y. Hurol, D. Urban Vestbro and N. Wilkinson (Eds.), *Methodologies in Housing Research*. Gateshead, Tyne and Wear: The Urban International Press.

Donovan N. and Halpern, D. (2002).Life satisfaction: the state of knowledge and implications for government, London: Cabinet Office/Prime Minister's Strategy Unit.
Garreau, J. (1991). *Edge City: Life on the New Frontier*, London: Doubleday.
Gärling, T. and Hartig, T. (2000). 'Environmental psychology and the environmental (design) professions', *Newsletter of the International Association of Applied Psychology*, 121, 1, 30-32.
Gifford, R. (1998). Special places. *Journal of Environmental Psychology*, 18 (1), 3-4.
Horelli, L. (2004). Enquiry by Participatory Planning within Housing. In Y. Hurol, D. Urban Vestbro and N. Wilkinson (Eds.), *Methodologies in Housing Research*. Gateshead, Tyne and Wear: The Urban International Press.
Hurol, Y., Urban Vestbro, D. and Wilkinson, N. (Eds.) (2004), *Methodologies in Housing Research*. Gateshead, Tyne and Wear, UK: The Urban International Press.
Jackson, T. (2002). *Quality of Life, Economic Growth and Sustainability*. In M. Cahill and A. Fitzpatrick (eds.) Environment and Welfare: towards a green social policy, Palgrave Macmillan, London, 97-116.
Kellet, P. (2004). Exploring Space: Researching the use of domestic space for income generation in developing countries. In Y. Hurol, D. Urban Vestbro and N. Wilkinson (Eds.), *Methodologies in Housing Research*. Gateshead, Tyne and Wear: The Urban International Press.
Lawrence, R.J. (1987). *Housing, Dwellings and Homes. Design Theory, Research and Practice*. London: John Wiley and Sons Ltd.
Marans, R.W. (2004). Modelling residential quality using subjective and objective measures. In Y. Hurol, D. Urban Vestbro and N. Wilkinson (Eds.), *Methodologies in Housing Research*. Gateshead, Tyne and Wear: The Urban International Press.
Marans, R.W. and Couper, M. (2000). Measuring the quality of community life: A program for longitudinal and comparative international research. Paper presented at the *Second International Conference on Quality of Life in Cities*, Singapore, March 2000.
Moser, G. (2000). Applying general psychology or doing environmental psychology, *Newsletter of the International Association of Applied Psychology*, 121, 1, 34-35.
Romice, O. and Uzzell, D. (forthcoming). Forty Years On: A Capital Experience in Psychology/Architecture Collaboration.
Proshansky, H.M., Ittelson, W.H. and Rivlin, L.G. (1970). *Environmental psychology: man and his physical setting*. New York: Holt, Rinehart and Winston.
Rowe, P. G. (1991) *Making a Middle Landscape*, Cambridge, Mass: MIT Press.
Symes, M. (1984). Learning from design. In J.A. Powell, I. Cooper and S. Lera (Eds.): *Designing for Building Utilisation* (pp. 199-205). New York: E. and F.N. Spon.
Turgut, H. and Kellet, P. (Eds.) (2001). *Cultural and Spatial Diversity in the Urban Environment*. Istanbul: YEM Yayin.
Uzzell, D. (1995). 'The Myth of the Indoor City', *Journal of Environmental Psychology*, 15, 4, 299-310.
Uzzell, D. (2000). Environmental psychology and the environmental (design) professions - a comment on Gärling and Hartig, *Newsletter of the International Association of Applied Psychology*, 12, 1, 32-34.

Chapter 2

The Perception of Urban Space from Two Different Viewpoints: Pedestrians and Automobile Passengers

Ricardo García-Mira and Myriam Goluboff

Introduction

The last few decades have seen a progressive revitalization of our cities, bringing in its wake the rehabilitation of a large number of public spaces that contribute to the leisure and welfare of their inhabitants. This transformation of urban space, however, has had to compete on the one hand with the powerful forces of land speculation, and on the other with a dramatic increase in the number of vehicles on the roads and streets; both of which pose major problem for local government. The habits that have been developed by adult automobile users have been reinforced on two levels: firstly, by the status associated with the possession and use of a private vehicle, and secondly, by the greater safety associated with the transport of children and adults alike by automobile, an association which is partly due to parental fears for their children's safety (Harden, 2000; Francis and Lorenzo, 2002; Björklid, 2002). Other factors that have led to an increased use of the automobile for journeys within the city are the 'adultization' of childhood as a result of the need to occupy children's time with activities that are programmed for them by adults, such as music lessons or sport (Francis and Lorenzo, 2002; Risotto, 2002), or parental belief that there is no other alternative, that they live a long way from schools or colleges, or that it is more convenient to take them by automobile on the way to work (Gatersleben, 2002). As a result, the disconnection between children's experience and the urban environment, and their being deprived of the ability to freely explore the environment for themselves (see also Rissotto and Tonucci, 2002), with the resulting logical prejudice to journeys on foot, has had a clear effect on the acquisition of knowledge of the environment and on the opportunity to experience the environment and integrate it cognitively.

The increased use of the automobile in our cities and the sharp rise in the number of vehicles on the road have meant a considerable reduction in opportunities for children to explore their environment, and have transformed urban space into a rather dangerous place for children to inhabit. This constitutes a major barrier to the overall aim of designing sustainable cities in which children are able to develop spatial awareness and sufficient knowledge of their environment.

It should be pointed out that spatial knowledge of one's environment has been studied from many different points of view, such as moving from one place to another, the use of maps or photographs, verbal descriptions, and more recently, designing virtual environments (Jansen-Osmann and Berendt, 2002). In a previous pilot study, a process based on estimating distances showed that children who move around their city by automobile do not appreciate their environment as a spatial continuum, but rather as a series of independent spaces that are reached by automobile or bus, thereby evidencing a different way of conceptualizing urban space in the light of different cognitive structures (Goluboff et al., 2002).

The present study is therefore concerned with the process of comparing how boys and girls who travel around their city by automobile on the one hand, and on foot on the other, gain an understanding and knowledge of urban space, and contrasting the cognitive structure that the different groups give to a series of easily identified places in their home town. The study also analyses the way in which boys and girls construct their cognitive maps. We follow the cognitive map concept from Downs and Stea (1977); according to these authors, a cognitive map is a product, a person's organized representation of some part of the spatial environment. As Stea et al. (1997) have pointed out, human behavior in the physical environment is an essentially spatial behavior consisting in the drawing up of cognitive maps which are vital for adding to our store of knowledge of our environment and for guiding the actions we take. This ability to build cognitive maps is something which is developed in the early stages of a child's development (see Blades et al., 1998; Stea et al., 2002).

The principal aim of this study is, therefore, to analyze differential aspects of the way in which boys and girls of different ages conceptualize urban space when they move around the city by automobile or on foot. The study also discusses the implications that this may have for urban planning, and the important role played by spatial knowledge in spatial interaction and the organization and accessibility of public services in our cities.

Methods

Sample

The initial sample consisted of 100 primary and secondary school students of between 10 and 17 years of age; 50 percent of the sample were male.

The sample was later reduced to 78 schoolchildren, since 22 protocols had to be eliminated because they had been wrongly or partially completed, mainly by the younger schoolchildren.

Procedure

The study was designed to confirm our hypothesis by asking the subjects to participate in two different exercises: a) Drawing a personal map on a sheet of paper in which they described as much as they could remember of their journey from home to school, and then asking them to write on the other side of the sheet of paper the name of thirteen places in the city that they could remember; b) The most frequently recurring place-names were chosen as the basis for the second exercise, which consisted in constructing a matrix with empty cells which the students had to fill in with the perceived straight-line distances that they thought existed between each place and each of the others. By this means they were asked to consider the distance between places without taking into account the way in which they traveled between them, in other words as if they were traveling in a straight line.

Analysis of the Data

For the first exercise, the maps drawn by the schoolchildren were analyzed in detail. For the second, the data obtained from the subjects was used as the input for a multidimensional scaling analysis which produced the following symmetrical matrixes: 1) Subjects who travel to school on foot; 2) Subjects who travel to school by automobile; 3) Boys who travel to school on foot; 4) Girls who travel on foot; 5) Boys who travel by automobile; 6) Girls who travel by automobile; 7) Subjects of between 10-11 years of age who travel on foot; 8) Subjects of between 10-11 years of age who travel by automobile; 9) Subjects of between 12-13 years of age who travel on foot; 10) Subjects of between 12-13 years of age who travel by automobile; 11) Subjects of between 16-17 years of age who travel on foot; 12) Subjects of between 16-17 years of age who travel by automobile. A further matrix was generated incorporating the real distances between places taken from a map of the city of A Coruña; this matrix was also used as an input for multidimensional scaling. The theoretical framework underlying multidimensional scaling is fully described in the literature (see Carroll and Arabie, 1980; Wish and Carroll, 1974).

Calculations were made of the correlations between the coordinates for the dimensions corresponding to real distances and those of the dimensions corresponding to perceived distances.

Results

Analysis of the Drawings Expressing Subjects' Personal Maps of Their Journey to School

Schoolchildren who walk to school have a more detailed awareness of urban space. They reflect the presence of shops, trees, friends' houses and in particular, an enlarged representation of pedestrian crossings, thereby revealing their fear when a pedestrian route coincides with a route used by motor vehicles. Schoolchildren who travel by automobile show roads and streets as a continuum with occasional buildings or elements of the urban landscape, sometimes labeled. It should be mentioned that the method used to analyze the drawings is more effective in the 10-11 and 12-13 age groups, since the 16-17 age group trivialises this kind of expression. No differences were found between boys and girls.

Analysis of the Complete Sample

The representation of real straight-line distances between the places used to calculate the estimated distances allows us to see that Dimension 1 (DIM-1) corresponds to the real routes in a transversal direction compared to the linear routes on either side of the isthmus – the beaches on ones side, and the harbor area on the other – along which the majority of the places included in our sample lie. Dimension 2 (DIM-2) corresponds to the longitudinal routes (see Figure 2.1). A comparison of the estimated distances and the real distances showed some specific features that will be commented on below.

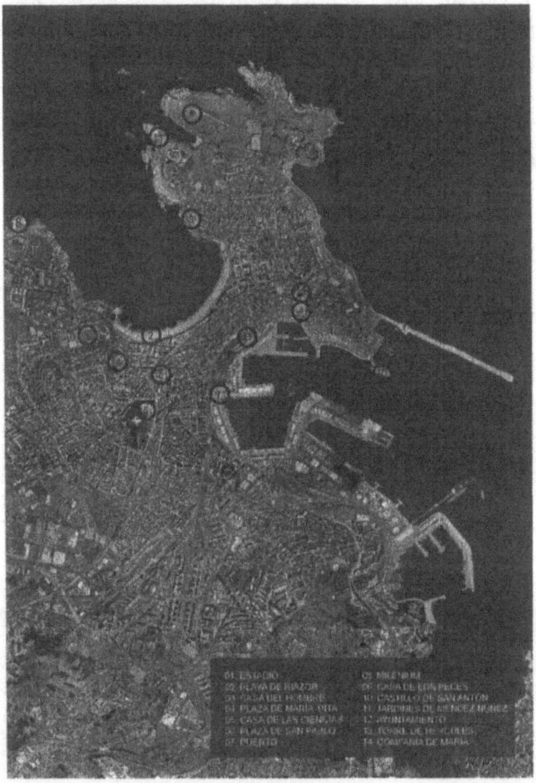

Figure 2.1 Real map and selected places

Analysis of the cognitive maps of boys and girls travelling on foot or by automobile The map corresponding to the boys and girls who walk to school showed a great deal of similarity to the real map, as can be seen by the fact that places are located in the same quadrant and distances between them are similar (see Figure 2.2, left).

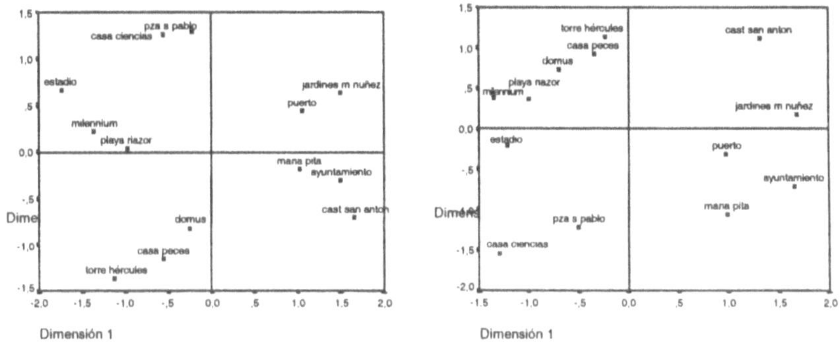

Figure 2.2 Representation of places in space for boys and girls who travel to school on foot (left) or by automobile (right)

In the case of children who travel by automobile, however, (see Figure 2.2, right), their cognitive map showed a different configuration of places when compared to the map of real distances. An interesting feature regarding the places located along the western side of the city, which corresponds to the line of beaches, is that distances are smaller than the real ones, and all the places located along the beachside appear in the same quadrant.

Analysis of the correlations between dimensions This analysis shows that the DIM-1 corresponding to children who travel on foot correlates (r=0,747**) with the DIM-1 of the coordinates of the real data, and is greater than that between the latter and the DIM-1 of the children who travel by automobile (r= 0,589*). In the case of DIM-2, there is a correlation of r=0,841** between those who travel on foot and the real data, this being greater than – and of the opposite sign to – that between the latter and the DIM-2 of those who travel by automobile (r= -0,479*).

Analysis of the Sample According to Sex

Sub-sample of girls On comparing the maps of the girls who go to school on foot and the girls who go by automobile, it can be seen that although there is a certain degree of distortion in both cases, this was greater in the case of those who travel by automobile (see Figure 2.3, GIRLS).

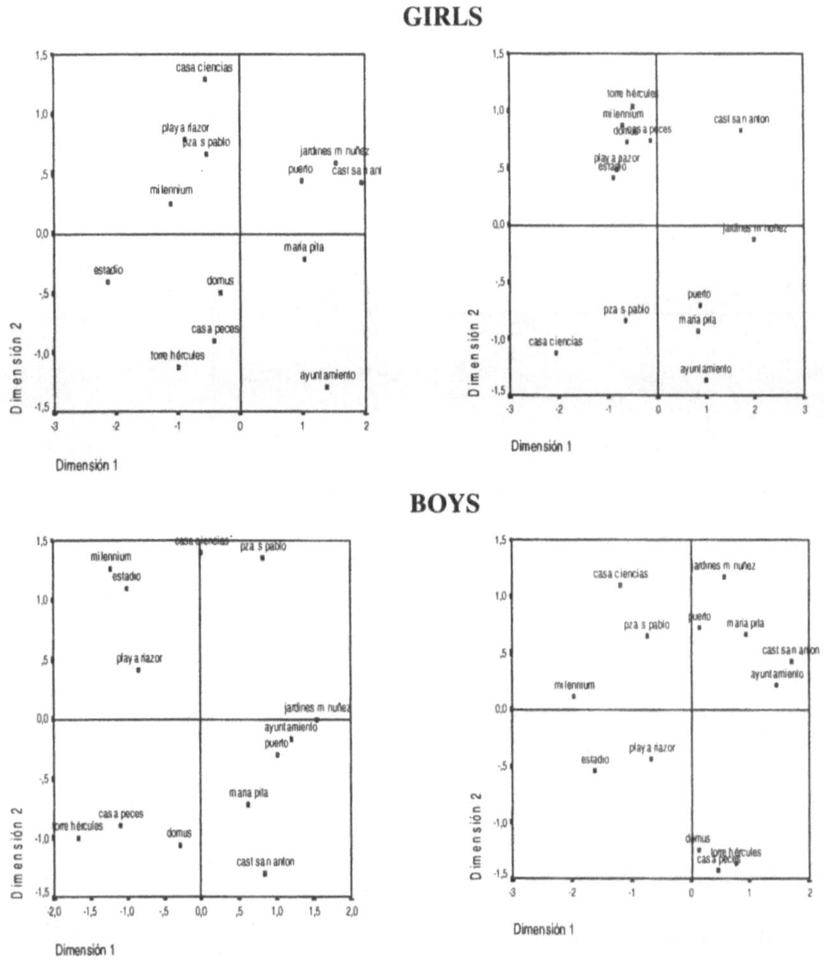

Figure 2.3 Representation of places in space in girls and boys who travel on foot (left) or by automobile (right)

Sub-sample of boys If we compare the maps of boys who go to school on foot with those of boys who go by automobile we can see that the distances estimated by the former are closer to the real distances than those estimated by the latter, for whom the different places appear outside their corresponding quadrants, thus producing a representation that differs greatly from the real one (see Figure 2.3, BOYS).

Analysis of the correlations between dimensions In the sub-sample of girls certain differences can be observed, such as the fact that the DIM-1 of those who travel on foot shows a significant correlation with the same dimension of the real data (r=0,695**), this correlation being greater than that between the real data and the DIM-1 of those who travel by automobile (r=0,481). In the case of the second dimension, there is also a significant correlation (r=0,647*) between the girls who travel on foot and the real data, again greater than that between the DIM-2 of the girls who travel by automobile and the real data for the same dimension (r=-0,205).

With regard to the sub-sample of boys, the DIM-1 of those who travel on foot showed a significant correlation (r=0,849**) with the DIM-1 for the real data, this correlation being greater than that between the DIM-1 of the boys who travel by automobile and the real data (r=0,370). In the case of the DIM-2 for those who walk to school, there is a correlation of r=0,928** with the DIM-2 of the real data, and this correlation is also greater than – and of the opposite sign to – that between the real data for DIM-2 and the same dimension for those boys who go to school by automobile (r=-0,363).

Analysis of the Sample by Age Group

Analysis of cognitive maps In the case of the youngest children (10-11 years old), the analysis of the maps of the two subgroups ('pedestrians' and 'passengers') enables us to see that there is some distortion in both cases. Nevertheless, whilst the boys and girls who travel on foot evidence a more disperse distribution of places, those who travel by automobile showed a greater degree of concentration of places around the axis of the ordinates, thus producing a greater degree of difference with respect to the real data. Similar differences were found for the 12-13 year-olds. Finally, the 16-17 year-old group showed a greater inconsistency in the data, since the configuration corresponding to those who go to school on foot reflects a rotation with regard to the axes of distribution, the surface area of the city appearing greater than in the real configuration, whilst for those who go to school by automobile the tendency to structure places in conglomerate groupings persists, following the trend that prevails in the analysis of the complete sample.

Analysis of correlations between dimensions by age group The greatest differences between travelling by automobile and travelling on foot are to be found in the youngest children (10-11 year-olds), the correlation between DIM-1 for the real data and the same dimension for the children who go to school on foot being r=0,636*, whilst in the case of those who go by automobile the corresponding correlation is r= 0,337. With regard to DIM-2, there is no statistical significance in the correlation between the real data and the data for either of the two groups of children in this age group (r=0,064 for those who go on foot and r=0,376 for those who travel by automobile). In the case of 12-13 year-olds and 16-17 year-olds, both those who go to school on foot and those who go by automobile show different degrees of correlation with the real data, with some inconsistencies being observed that are probably due to them finding tasks they were asked to perform somewhat boring. This indicates that we need to review the way in which our tasks are administered, and also to increase the size of our sample.

Conclusion

The results of this experiment provide us with information about two different experiences, one being that of spatial awareness as perceived from the inside of a private motor vehicle, and the other that of the same spatial awareness as perceived by people as they walk to school. The mental representation of space and the way in which the boys and girls in question understand their personal experience differs according to the mode of travel they use. The use of multidimensional scaling to generate these mental representations, in the form of cognitive maps, shows that there are different ways of conceptualizing the same space depending on the age of the children, their sex, and the mode of transport used. Thus, we observe that those children who walk to school conceptualize urban space in a different way to those who travel by automobile, perceiving the different places included in the survey as forming part of a continuum, remembering more items or objects and demonstrating a greater ability to structure space. On the other hand, the children who travel by automobile have a poorer perception of space, showing a tendency to group different places together, a lesser degree of structuring and an absence of any impression of continuity. In the case of this latter group, the places that correspond to a journey along the coast road appear as being much closer together than they really are (it may be assumed that traffic travels faster along this road, creating a perception of closeness between the places to be found along it). However, when we study the relationships between the places that have a real spatial relationship between one side and the other of the isthmus on which the city is located, in other words perpendicular to the coast line, the degree of distortion is similar for both male and female subjects who travel on foot and by automobile. It seems as if movement between these places in a motor vehicle is not direct, and therefore it is impossible either to have an overall perception or to reduce the journey time.

When the sample is analyzed from the point of view of the sex of the subjects, we find that the cognitive map of boys who walk to school is much closer to the map of the real data than is that of girls who walk to school. There is also a greater degree of distortion in the cognitive map of girls who travel to school by automobile, particularly in the concentration of places that appear as being close together, when in reality they are further apart. The boys who travel by automobile also show a certain degree of distortion as regards the estimation of distances, but these distances are nevertheless closer to the real ones.

When the differences are analyzed according to age, the best structured responses from those children who walk to school came from the 12-13 year-olds rather from the 10-11 year-olds, whilst in the case of the 16-17 year-olds better responses were obtained from those who travel by automobile. There is, however, another fact that should be taken into account if we are to understand the significance of a behavioral study of young people according to their age. In the case of 10-11 year-olds, the fact that they travel to school by automobile would seem to imply a greater dependence on adults in the use of this resource. 16-17 year-olds, on the other hand, are able to move freely about the city on foot as well as by automobile. For this reason, perhaps, the mode of transport to school should not be taken as the prime parameter in this case.

Our findings suggest that the city has to be experienced if we wish to promote a greater awareness of our surroundings; this will be the first step towards developing an attitude to the environment. If we consider this to be of importance, and that travelling around the city on foot is also important, then given the consequences for interaction and spatial cognition that derive from this, the way in which the city is organized should take this into account in terms of both distance and accessibility. These findings also suggest the need to set up participative programs for children in which they can learn to explore their surroundings as the result of a process of interaction (e.g. with teachers and town planners), encouraging them to take an active interest in their local environment and taking note of their initiatives in a participative context involving the whole community, children included.

References

Blades, M., Blaut, J.M., Darvizeh, Z., Elguea, S., Sowden, S., Soni, D., Spencer, C., Stea, D., Surajpaul, R. and Uttal, D. (1998). A cross-cultural study of young children's mapping abilities. *Transnational Institute of British Geographers*, 23, 269-277.

Björklid, P. (2002). Parental restrictions and children's independent mobility from the perspective of traffic environmental stress. In R. García-Mira, J.M. Sabucedo and J. Romay (Eds.): *Culture, Quality of Life and Globalization*. Proceedings of the 17th Conference of IAPS (pp. 762-763). A Coruña: AGEIP-IAPS.

Carroll, J.D., and Arabie, P. (1980). Multidimensional scaling. In M.R. Rosensweig and L.W. Porter (Eds.), *Annual Review of Psychology* (Vol. 31, pp. 607-649). Palo Alto, CA: Annual Reviews.

Downs, R.M. and Stea, D. (1977). *Maps in minds. Reflections on cognitive mapping*. New York: Harper and Row Publishers.

Francis, M. and Lorenzo, R. (2002). Seven realms of children's participation. *Journal of Environmental Psychology*, 22, 157-169.

Gatersleben, B. (2002). Sustainable school travel: what is the potential for change? In R. García-Mira, J.M. Sabucedo and J. Romay (Eds.): *Culture, Quality of Life and Globalization*. Proceedings of the 17th Conference of IAPS (pp. 762-763). A Coruña: AGEIP-IAPS.

Goluboff, M., García-Mira, R. and García-Fontán, C. (2002). Percepción del espacio urbano desde la perspectiva de peatones y pasajeros [Perception of urban space from pedestrian and passenger perspective]. In R. García-Mira, J.M. Sabucedo and J. Romay (Eds): *Psicología y Medio Ambiente. Aspectos Psicosociales, Educativos y Metodológicos [Psychology and Environment. Psychosocial, Educational, and Methodological Aspects]*, (pp. 149-157). A Coruña: AGEIP-IAPS.

Harden, J. (2000) There's no place like home: the public/private distinction in children's theorising of risk and safety. *Childhood*, 7, pp. 34-45.

Jansen-Osmann, P. and Berendt, B. (2002). Investigating distance knowledge using virtual environments. *Environment and Behavior*, 34 (2), 178-193.

Rissotto, A. (2002). Projects and policy for childhood in Italy. Paper presented at the II Congress of Environmental Psychology in Italy. La Sapientza – CNRS University, Italy.

Rissotto, A. and Tonucci, F. (2002). Freedom of movement and environmental knowledge in elementary school children. *Journal of Environmental Psychology*, 22, 65-77.

Stea, D., Elguea, S. and Blaut, J.M. (1997). Desarrollo del conocimiento del espacio a la escala macroambiental entre niños muy jóvenes. Una investigación transcultural [Development of space knowledge in a macro-environmental scale among very young children. A cross-cultural research], *Revista Interamericana de Psicología / Interamerican Journal of Psychology*, 31 (1), 141-147.

Stea, D., Elguea, S., LeFebre, M., Pinon, M., Blaut, J.M. (2002). Teoría ecológica, mapeo universal y cognición ambiental aplicada [Ecological theory, universal mapping and applied environmental cognition]. In R. García-Mira, J.M. Sabucedo and J. Romay (Eds): *Psicología y Medio Ambiente. Aspectos Psicosociales, Educativos y Metodológicos [Psychology and Environment. Psychosocial, Educational, and Methodological Aspects]*, (pp. 85-109). A Coruña: AGEIP-IAPS.

Wish, M. and Carroll, J.D. (1974). Applications of individual differences scaling to studies of human perception and judgment. In E.C. Carterette and M.P. Friedman (Eds.), *Handbook of perception* (Vol. 2, pp. 449-488). New York: Academic Press.

Chapter 3

Space Use, Dwelling Layout and Housing Quality: An Example of Low-Cost Housing in Istanbul

Ahsen Özsoy and Gülçin Pulat Gökmen

This chapter aims to discuss the problem of housing quality by examining the intangible aspects of space use in low-cost housing in Istanbul. As one of the most common building types in the built environment, housing is a fundamental and complex phenomenon. It has different meanings for different cultures, different groups and different individuals. Particularly in countries having a major housing problem, it is important to understand and to study varying aspects of housing.

In Turkey, layouts and plan organizations of various dwelling types show different changes that conform to the times. The most important feature of the traditional Turkish house is the 'room' with its various units serving different activities in the house. The plan organization of the house is based on the number and shape of the rooms. The other common element in the plan organization is the 'sofa' which is a circulation area and a place for social interaction. The different combinations and uses of the rooms and the shape of the 'sofa' present a large number of plan types (Kuban, 1995).

In the last 50 years, during the process of urbanization in Turkey, dwelling units that meet the housing needs of low-income population are being built both in planned and unplanned ways in big cities like Istanbul. Informal housing examples, i.e. squatter dwellings built by people migrating from rural areas to urban centers, make up the most important group of low-cost housing. Newcomers produce dwelling environments according to their lifestyles, their rural backgrounds and their preferences. Some even build on top of existing structures, adding storeys as family and economic demands necessitate. Phases of plan development and factors affecting plan layouts have been examined in various studies (Atalik et al., 1986).

Planned housing is built on both an individual and a cooperative basis. Dwellings constructed through cooperatives are the most important part of the total housing production. Because of the lack of housing in Turkey, qualitative aspects of dwellings have generally been neglected, and quantitative aspects have been emphasized. The concept of quality has only been considered for high-income housing in terms of location in the city, dwelling size, building materials, etc., but not in terms of dwelling layout or quality of space use.

Dwelling Quality, Dwelling Layout and User Satisfaction

Norberg-Schulz (1985) suggested that 'to dwell', in its broadest meaning, is to 'be at home' in an environment and a dwelling is 'more than having a roof over our head' (p. 7). He described the four modes of dwellings as natural, collective, public and private and their architectural levels as settlement, urban space, institution and house (p. 13).

The concept of 'quality' has very broad usage, encompassing a variety of meanings defined by researchers in almost all fields. Quality refers to distinguishing properties that promote a degree of excellence (Smith, Nelischer, and Perkins, 1997). Discussion of quality implies a search for criteria that may be used to distinguish between good and bad. Clearly, different criteria are likely to be important to different individuals, different social groups and different household types. However, it is also possible to define shared criteria that are more or less common to all residents. According to Goodchild (1997, p. 14), unraveling the concept of quality in housing brings up a series of questions. What are the main differences between types of housing and types of layout? What are the advantages and disadvantages of these types? What are the different aspects of quality? How can a designer or developer identify and measure perceptions of quality in different contexts? The first step is to distinguish between 'habitability' and what might be called the 'socio-cultural aspects' of quality. Most discussions of quality in housing design concentrate on its socio-cultural aspects. They focus on how people experience the environment around them; how they interact with that environment; and how they judge its suitability in relation to their daily routines and their expectations for the future (Goodchild, 1997, p. 32). The question is how to classify these less definable, often more variable qualities into different aspects. Rapoport (1989) has proposed an 'environmental quality profile' as a way of trying to describe and communicate the many components of the concept of quality, and the nature, ranking, importance or magnitude of these components. He has also emphasized the importance of perceived quality and the subjectivity of the concept.

Lawrence (1995) stated that the concept of housing quality required a reappraisal, and that there was an urgent need for an integrated definition of housing quality in which sets of architectural, demographic, economic, ecological and political factors are explicitly interrelated.

Despite the polyvalence of human goals and values, in recent years some authors have proposed a hierarchy of needs (Lawrence, 1987, p. 159). For example Cooper (1975) re-organized Maslow's need system into a simple hierarchy: i. shelter, ii. security, iii. comfort, iv. socialization and self-expression, v. aesthetics. In fact, the housing 'needs' of a resident, or of a household, do not correspond to a constant norm but are defined by a complex matrix of interrelated factors that change over the course of time (Lawrence, 1989). Housing settlements with very stereotyped plans that are designed for a large number of families without considering their opinion in the design process are far from meeting the users' changing needs in different stages of their dynamic life cycles. These stereotyped dwelling units inevitably undergo certain changes and alterations in time (Özsoy and Esin, 1988) both in terms of internal and external space.

According to Francescato (1993), use was the original reason for building houses. The term has often been taken to mean engaging in activities. In planning and architecture it tends to be synonymous with 'function'. He claimed that it seemed more appropriate to think of use as any interaction of people with their residential environment (p. 42). Use gives meaning to housing, and at the same time meaning guides how housing is used and these are the two most important topics to housing policy, planning or design and quality in any cultural context around the world (Arias, 1993).

There is a considerable amount of research on the arrangement of domestic space as a quality factor revealing the social identity of its dwellers. Some researchers consider the relationship between analysis patterns of furniture and the socio-economic status of the residents. For instance, Laumann and House (1970) identified four distinct types of interior decoration associated with social classes characterized by both income and social life style. They claimed that living room decoration and furnishings reflect their social identity. In their view, residents decorate their home according to their cultural backgrounds and values. For example, less educated people tend to prefer traditional furniture whereas the more educated ones favor a more informal style of interior decoration (Laumann and House, 1970).

As Lawrence pointed out (Lawrence, 2000, p. 58), there were very few studies that include morphological or spatial dimensions of housing. He also argued that it was impossible to identify changes in the design and use of housing units over time. An integrative perspective with more collaboration, conceptual innovation, and multiple methods and measures were necessary.

Bernard, Bonnes and Giuliani (1993) compared pictures of living rooms of three European countries. The reason the room was chosen was that it is the privileged area of 'stage setting' where the dweller proposes to show an image of himself and that which he wishes others to have of him. They found that income groups differentiate in favor of social determinism, in aesthetic choices, in furnishing and decorating the habitat (Bernard, Bonnes and Giuliani, 1993).

Qualitative aspects of the housing environment have gained great interest in recent housing research in Turkey. Studies have shifted from environmental performance to psycho-social and spatial quality characteristics which can be measured by users' subjective evaluations (Bayazıt, Dülgeroğlu, Yılmaz and Çıracı, 1986; Dülgeroğlu, Aydınlı and Pulat, 1996).

In this study, the definition of 'quality' has been accepted as 'fitness for use'. Quality for housing can be evaluated by examining user satisfaction together with the environment. Important components for the study of user satisfaction are both physical characteristics of dwelling spaces, such as location in the city, size of dwelling, number of rooms, etc. and the subjective evaluations of users of their dwellings and environments. Layout is one of the most important characteristics that determine the quality for housing units. Space use and furniture arrangements, in conjunction with the dwelling layout, are considered as two interrelated indicators of housing quality. In this study, an example of low-income housing in Istanbul is examined in terms of the satisfaction of users with the plan layout of their individual dwellings.

Case Study

Method

The findings of the paper are based on the results of a research study (Özsoy, et al., 1995) carried out in three housing settlements in Istanbul having different socio-economic structures with the aim of making a qualitative assessment. In this study, we focus on one of these three settlements by analyzing the data collected from the interviews with the users and the observations of space use to determine the furniture arrangements and the alterations made in the dwellings. The area studied is one of the most interesting housing developments on the Anatolian side of Istanbul (Maltepe, Esenkent). In the 1970s, the district was planned as a squatter prevention area by the government, and the land was allocated to the low-income families who had been living in the squatter settlements near the chosen area. In the 1980s, it was decided that the unsettled part of the area was to be used for the cooperative housing of various institutions, firms and organizations (Figure 3.1).

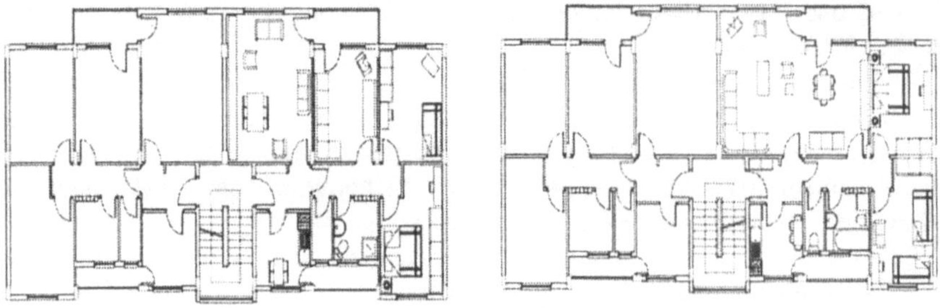

Original Plan **Transformed Plan**

Figure 3.1 Examples of original and transformed plans from Type A plan

The methodology of the field study was mainly based on the techniques of interview, observation and archival data. Interviews were conducted in the dwellings randomly chosen from apartment blocks in the settlement. A pair of graduate students who had been trained in the interview process conducted the interviews, filled out the questionnaires and made drawings of the changes and alterations people had affected in their dwellings. The drawings were made on the scaled plans that had been prepared along with the questionnaire form. They also drew the fixtures and furniture in the dwellings and took photographs with permission.

The plan layouts of different cooperatives, the socio-economic status of the users, and the space-use of dwellings are some of the topics that have been examined to make a comprehensive evaluation about the quality of life of the residents. The different plan layouts and potential of spatial organizations for different family structures are evaluated as an indicator of the spatial quality of dwellings. The relationship between the use of space and satisfaction from plan layouts is evaluated based on the data collected by using the techniques of interview and observation. 79 dwelling units from three different co-operatives have been examined in detail by analyzing the questionnaires, drawings and photographs.

Family Characteristics

In the study, the distribution of the families sampled according to family structure has been examined. The percentage of nuclear families is high (74.7%). Families with 3 or 4 persons are in the majority (61.4%). The second largest group is the 5-person family with 16.9% and the percentage of extended families is 12.0%. Approximately half of the inhabitants interviewed fall into two age categories: 30-39 and 40-49. Because the questionnaires were filled out in the dwellings, those interviewed were generally women (73.5%). As for the professions of the respondents, they were mostly housewives (47.0%), followed by white collar workers (15.7%), students (13.3%) and retirees (12.0%). About half of the people interviewed have a high school education (50.6%). The percentage of people having primary school education is 27.7%, while those with a university education is 20.5%.

Determination of income level for this type of study is generally difficult and carries the risk of not being aware of the real life conditions of the families. Therefore, as an indicator of income level, ownership of consumer goods has been examined and three categories have been observed according to the income level: the first group is the families having a refrigerator, washing machine, vacuum cleaner and color TV; the second group is the families having a tape recorder, radio and music set in addition to the consumer goods of the first group; and the third group of families also has an automobile, dishwasher, video, and computer.

As for years spent in the dwelling, 61.4% of the families have lived in their homes for 6-10 years; the residents in the settlement have spent enough time to make a realistic evaluation about the quality of their dwelling and its environment.

Plan Types and their Uses

Three main plan types from the three different cooperatives examined in the settlement have similar block size and form. Two of them have 2 bedrooms and the third one has 3 bedrooms. All of the plan types have a living room, a kitchen and a bathroom with a separate WC. As can be seen from Figures 3.2, 3.3, and 3.4, general layouts and spatial organizations of the plan types are basically similar. As one of the indicators of quality for dwellings, satisfaction with the spatial organization was asked. 67.5% of the interviewees responded positively about their dwellings, while 32.5% were not satisfied. The reasons for the dissatisfaction were an insufficient number of rooms (25.3%) and the size of the dwelling (4.8%). When they were asked what they thought about the qualities their dwellings would need to have, 28.9% emphasized a sufficient number of rooms, while 3.6% indicated sufficient dwelling size. Changes and alterations made by the families and arrangements of furniture showing various uses of space have been examined and grouped according to the plan types (A, B, C).

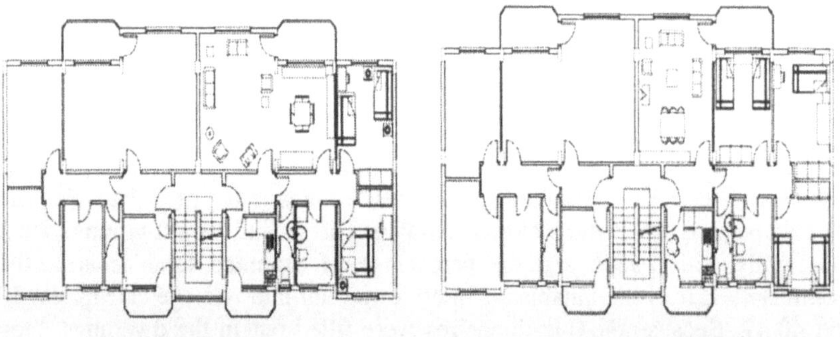

Original Plan **Transformed Plan**

Figure 3.2 Examples of original and transformed plans from Type B plan

Original Plan **Transformed Plan**

Figure 3.3 Examples of original and transformed plans from Type C plan

Figure 3.4 A general view from the settlement

Variations in the use of space and activity patterns occurring in the dwellings can be found in Table 3.1. In 73% of the dwellings, the living room is used only for visitors. To allocate a room for visitors is one of the customs of traditional Turkish families that is gradually disappearing in the urban lifestyle. This entails keeping one of the rooms in the dwelling clean and orderly. Studies conducted with the various income groups have shown the changing habits of the families in the urban areas and found a growing tendency to lose the traditional way of life in the urbanization process.

Table 3.1 Use of space and activities occurring in the rooms (total sample is 79)

Rooms	Activities Occurring In Rooms	Number of Dwellings
	Sitting, dining for guests	35
	Sitting, hosting guests	25
Living Room	Sitting, living, watching TV	4
	Hosting, watching TV	9
	Sitting, dining, watching TV	2
	Sitting, hosting, sleeping	2
	Sitting, watching TV, sleeping	1
	Sitting, sleeping	1
	Sleeping of parents	62
	Sleeping of parents, sleeping of children	6
Bedroom-1	Sitting	4
	Sleeping	4
	Sleeping of guests	2
	Sleeping, studying	1
	Ironing	1

Bedroom-2	Sleeping of children	26
	Sleeping of children, daily sitting	11
	Sleeping of children, studying	7
	Daily sitting, watching TV	5
	Sitting, sleeping of guests	5
	Dining	4
	Sitting	3
	Sleeping of children, of grandparents	3
	Sleeping	3
	Sleeping of children, sitting of guests	2
	Sleeping of parents	2
	Sleeping, studying	1
	Sleeping of children, dining	1
Bedroom-3	Sleeping of children	12
	Sleeping of parents	6
	Sleeping	3
	Sitting, watching, sleeping of children	2
	Sitting, sleeping	2
	Sitting	1
	Sewing, watching TV	1
	Watching TV, playing of children	1

As for the furniture used, there is sitting space for a total of 7 people, a dining table, a buffet, a TV and coffee tables in most of the dwellings (66%). In 24 dwellings (29%), there are two types of sitting areas (Figure 5). One of the commonly observed pieces of furniture in the dwellings of low- and middle-income groups is the convertible sofa which can be used normally for sitting and for sleeping when it is opened. As one of the customs of the traditional Turkish family, hospitality for guests who stay overnight is now less frequent with the changing lifestyles of urban families. Nevertheless, the sofa bed and the fold-out bed-type of furniture can be used in a flexible way to meet the needs of the families when they have overnight visitors.

One of the bedrooms is used as the 'master bedroom' by 82% of the families. While 32% use the second bedroom for the activities of children, 58% use it in a flexible way for daily activities (living, dining, watching TV, housework) and also for children's activities. For 88% of the families with one or more children, it is very clearly seen that most of the children have no separate room allocated for their own activities. It seems a contradiction for space use to keep one of the rooms for the guests while sharing one of the bedrooms with the children's activities.

Changes and alterations made to improve the quality of building layout.

The responses concerning the changes and alterations they had already made or planned to make in their dwellings can be grouped as changes and alterations for: improvement of space use; security of the dwelling; maintenance and repair; and a betterment of aesthetic quality.

Figure 3.5 Furniture arrangements in a living room

25.3% of the respondents said they had divided their living room to create additional space, 21.7% said they had not made any alterations or changes as yet. When they were asked about their intention to make alterations if they had the financial resources, 44.6% of the respondents stated they would make alterations, 43.4% said they would not. Types of alterations and changes they planned to make were to divide the living room into two separate areas (19.3%), and to add a balcony to the living room (12.9%).

Type A The number of questionnaires related to dwellings with the Type A plan is 31. Among these dwellings, the original plan with three bedrooms was found in 7 homes only. The living room is rectangular and used for multiple purposes such as entertaining guests, dining and relaxing. In general, one of the three bedrooms is used for daily life. In 10 dwellings (32%), the wall between the living room and bedroom has been removed and the living room has been enlarged into an 'L' shape. In this type of dwelling, there are two groups of furniture for seating, creating an opportunity for dining, watching TV and sleeping if necessary. In 8 dwellings (26%), the wall between the living room and bedroom has been removed, but folding panels have been installed to be used when needed. In these examples, an extra room for guests is created by closing the panels. Another use of this moveable wall is to host different groups at the same time. For example, the guests of parents and the guests of teenage children can be welcomed simultaneously. In one dwelling, the living room and bedroom are separated by a buffet after removing the wall between them, and this newly created room is used as a bedroom. However since the separator is flexible, this room can be used as one large space for other purposes as well (Figure 3.2).

Type B The number of questionnaires with the Type B plan is 20. 11 of these dwellings (55%) that have original plans with two bedrooms have been changed. The living room is in the shape of an 'L' and used for different purposes such as entertaining guests, dining and watching TV. In 7 homes (35%) with the original plan, there are two types of sitting room furniture. One of these groups is utilized for daily living while the other is kept for visitors. Two houses were changed: in one of them the living room is divided by a wall; in the other, the living room is divided by a piece of furniture. The former is used as a bedroom, while the latter is used as a daily living room (Figure 3.3).

Type C Type C is similar to Type B with two bedrooms. The number of questionnaires conducted in Type C is 28. The original plan has been used without change. The furniture in the living room carries out functions such as relaxing, hosting, watching TV; and a table and buffet are used for dining with guests. In 7 dwellings with their original plan, there are two sets of furniture for sitting. One of these is used for daily sitting and watching TV. In 6 dwellings, the living room is divided as in the Type B plan. In two houses, this division is achieved with a wall, while it is achieved using furniture in four dwellings. The extra room is used as a bedroom in three dwellings, and a daily living room and bedroom in the other (Figure 3.5).

Discussion

Quality of a dwelling space is related with the potential to meet the needs of the users. In this study, the definition of 'quality' has been accepted as 'fitness for use'. Variety and richness of the activities that take place in the dwelling unit are the qualitative indicators of the plan layout. Space use and furniture arrangements, related with the dwelling layout, are considered as two interrelated indicators of housing quality.

General layouts and spatial organizations of the plan types examined in the study are basically similar. The user group studied in this chapter consisted of 3-4 person nuclear families from middle- and low-income groups, generally having a high school education. Their qualitative evaluations show their growing consciousness of the qualities of the environment and the dwelling itself. They mostly evaluate the plan configuration positively. The families who are not satisfied with their dwellings have developed some solutions to adapt the dwellings to their lifestyles and family structures. Dividing the living room to create an additional room, or removing the wall between the living room and the bedroom to obtain a larger and more flexible living area are some of these examples. In addition to the structural solutions, people prefer to use multi-purpose furniture to increase the flexibility of the rooms. These findings also show the changing habits of the families in the urban areas with a growing tendency to lose the traditional way of living in the urbanization process.

Because of common problems observed in housing settlements that are produced without considering users in the design process, it is very important to conduct qualitative evaluations to examine how and to what extent they meet users' changing needs in different stages of their dynamic life cycles. The findings of the study indicate the meaningful touches of the individual users in their homes that increase their satisfaction and show the tendencies and preferences which shed some light for future design activities of mass housing.

References

Arias, E. G. (1993). User Group Preferences and Their Intensity: The Impacts of Residential Design. In E. G. Arias (ed.), *The Meaning and Use of Housing International Perspectives, Approaches and Their Applications* (pp. 169-199). Aldershot: Ashgate.

Atalık, G., Bölen, F., Ünügür, M., Aksoylu, Y., Özsoy, A., Dener, A. and Köksüz, B. (eds.). (1986). *Field Studies in the Squatter Settlements of Istanbul*. Istanbul: Istanbul Technical University, Faculty of Architecture.

Bayazıt, N., Dülgeroğlu, Y., Yılmaz, Z. and Çıracı, M. (1986). *Toplu Konutlarda Mekan Standartları Araştırması* [Research on Spatial Standards in Mass-Housing]. Istanbul: Istanbul Teknik Üniversitesi, Mimarlık Fakültesi.

Bernard, Y., Bonnes, M. and Giuliani, M. V. (1993). The Interior Use of Home: Behavior Principles Across and Within European Cultures. In E. G. Arias (ed.), *The Meaning and Use of Housing International Perspectives, Approaches and Their Applications* (pp. 81-101). Aldershot: Ashgate.

Cooper, C. M. (1975), *Easter Hill Village: Some social implications of design*. NewYork: Free Press.

Dülgeroğlu, Y., Aydınlı, S. and Pulat, G. (1996). *Toplu Konutlarda Nitelik Sorunu* [Quality Problem in Mass Housing]. Ankara: T. C. Başbakanlık Toplu Konut İdaresi.

Francescato, G. (1993). Meaning and Use: A Conceptual Basis. In E. G. Arias (ed.), *The Meaning and Use of Housing International Perspectives, Approaches and Their Applications* (pp. 35-50). Aldershot: Ashgate.

Goodchild, B. (1997). *Housing and the Urban Environment, A guide to housing design, renewal and urban planning*. Oxford: Blackwell Science Ltd.

Kuban, D. (1995). *The Turkish Heat House*. Istanbul: Eren Press.

Laumann, E. O. and House, J. S. (1970). Living Room Styles and Social Attributes: The Patterning of Material Artifacts in a Modern Urban Community. In E. O. Laumann, P. M. Siegel and R. W. Hodge (eds.), *The Logic of Social Hierarchies*, Chicago: Markham.

Lawrence, R. J. (1987). *Housing, Dwellings and Home: Design Theory, Research and Practice*. Chichester: John Wiley and Sons.

Lawrence, R. J. (1989). Public and Private Responsibilities in the Definition of Housing Quality. In N. Wilkinson (ed.), *Quality in the Built Environment, Conference Proceedings* (pp. 123-129). Newcastle Upon Tyne: The Urban International Press.

Lawrence, R. J. (1995). Housing Quality: An Agenda for Research. *Urban Studies*, 32 (10), 1655-1664.

Lawrence, R. J. (2000). House Form and Culture: What have we Learnt in Thirty Years?, in K. D. Moore (ed.), *Culture-Meaning-Architecture, Critical reflections on the work of Amos Rapoport* (pp. 53-76). Aldershot: Ashgate.

Norberg-Schulz, C. (1985). *The Concept of Dwelling: On the way toward figurative architecture*. NewYork: Electa/Rizzoli.

Özsoy, A., Esin-Altaş, N., Ok, V. and Pulat, G. (1995). *Toplu Konutlarda Davranışsal Verilere Dayalı Nitelik Değerlendirilmesi*. [Quality Assessment in Mass-Housing Based on Behavioral Data]. (Research Report TÜBİTAK-İNTAG 102). Istanbul: Istanbul Technical University, Faculty of Architecture.

Özsoy, A. and Esin, N. (1988). *Toplu Konutlarda Tasarım, Yapım Sistemi ve Mekan Kullanımı Etkileşiminin Araştırılması*. [Interaction Between Design, Construction

System and Space-use in Mass-housing]. (Research Report) Istanbul: Istanbul Technical University, Environment and Urban Planning Research Center.

Rapoport, A. (1989). Environmental Quality and Environmental Quality Profiles. In N. Wilkinson (ed.), *Quality in the Built Environment, Conference Proceedings* (pp. 75-83). Newcastle Upon Tyne: The Urban International Press.

Smith, T., Nelischer, M. and Perkins, N. (1997). Quality of an urban community: a framework for understanding the relationship between quality and physical form. *Landscape and Urban Planning, 39*, 229-241.

Chapter 4

An Evaluation of 'the Feeling of Security' in a New Mass Housing Compound in Istanbul

Suat Apak, Gokhan Ulken and Alper Unlu

The Definition of Crime

The definition of crime always forces us to take into account 'the feeling (or feelings) of security'. This concept is not only related to the occurrence of crime, but can also be explained as anxiety due to the existence and possibility of crime. Indeed, this anxiety can be defined as the fear of crime, rather than the physical and psychological effects of the crime itself; this notion first appeared as a separate issue at the beginning of the 1970s (Taylor, 1988). This emotional outcome always occurs before the actual occurrence of crime and depends predominantly on people's responses; ultimately, however, it affects social relationships, fully destroying what can be considered some of the basic qualities of life. The British Crime Survey argued in 1996 that measuring people's fear of crime is imperative as it acts as an indicator of the crime problem and of the degree of public concern regarding crime (Lawson and Heaton, 1999).

The notion of the fear of crime is an important social problem as it is a result of the emergence of feelings of insecurity. Fear of crime is a very common issue in urban areas, thus it creates what can be called more vulnerable populations. Going out at night and preferring secure places, and conversely avoiding potential crime areas, are some socio-behavioral phenomena which have caused a decrease in social interaction, social control and healthy neighbor relations.

Many propositions regarding the fear of crime have been put forward. Hunter claims that social disorganization results in the emergence of a fear of crime and a consequent decrease in social sensitivity. Wilson and Kelling also take up the same position, arguing that the notion of crime and disorganization in society leads to a decrease in social control (Taylor, 1987). These two views, although they have conceptual differences, accept the notion that the fear of crime is a highly socio-behavioral phenomenon. The main difference can be attributed to Hunter's view being primarily a psychological concept which is realized before the occurrence of the crime.

Many methods have been developed to decrease crime and to increase feelings of security in home environments, methods that aim to reconstruct individual security. This reconstruction of feelings of security is primarily linked to the avoidance of crime by the installation of mechanical devices, the reorganization of physical and social systems and by the realization of administrative regulations (Laymon, 1974; Marcus and Sarkissian, 1986).

According to recent technological developments, there are many devices that play an important role in crime prevention. Strategies like spatial planning, e.g. urban and architectural layouts, may be taken into consideration. Spatial layouts which lack transitionary spaces between public and private areas can also be said to be linked to the fear of crime; in other words, transitionary spaces can be defined as bumper zones between public and private spaces. According to this definition, the notion of crime is related to the condition of spaces in general – specifically behavioral settings and the temporal order of space.

With regard to the prominence of these issues, Newman (1973) suggests the concept of a defendable space that is a congruent model between the physical attributes and the social milieu of the area. This model intends to create housing clusters based on social relations, acknowledgement of physical attributes like roads and public areas, activity domains for user groups and opportunities for surveillance. Newman (1973) defines five design mechanisms that contribute to the concept of defensible space: the distribution of occupants, the territorial capacity of the physical environment, opportunities for surveillance, security potential in the environment and image.

Crime in Urban Areas

Crime prevention mechanisms in urban areas can be analyzed in accordance with the concept of categorization. These categorizations, such as the temporal, perceptive, ideological and economical, are taken into consideration when evaluating urban life and properties. Categorization of temporal and perceptive qualities of the physical environment are important issues in perceptions of security. These factors can be linked to spatial configurations like geometry, the structure of roads and the location of landmarks and boundaries. All these elements are highly significant in that they contribute to the meaning of the built environment (Unlu and Edgu, 2001). The concept of disorder in the physical structure and incongruity in its perception make citizens more insecure in and about their immediate surroundings. In these settings, citizens feel at risk and they may come to expect external damage from unforeseen stimuli (Apak, 1998). Consequently, the fear of crime is not only related to the quantitative aspects of life, but also to qualitative aspects of the environment.

The rapid increase of the population in Istanbul has created various problems for its urban areas, including issues such as insufficient shelter, (lack of) infrastructure and urban services, and physical and social deficiencies in the urban pattern. The fact is that Istanbul has not been able to integrate and assimilate newcomers into the social network, and this has resulted instead in a social duality within the city and in a decreasing social network and interaction in residential environments. On the outskirts of the city, unplanned, unauthorized, spontaneous settlements have occupied areas of land in the vicinity of industrial plants. The manufacturing industry and squatter areas have merged with one other, with squatters following the location of industry. In recent years, beyond the peripheral areas that were originally squatter settlements, newly planned residential developments have been built on the outskirts of the metropolis (Erkut, Ocakci and Unlu, 2001; Centel, 2000).

As result of unpredicted mobility, irregularities in structure and social heterogeneity have been observed in urban districts. Social heterogeneity has also appeared in historical neighborhoods of the city, and these neighborhoods recently presented descending population density and contrarily ascending crime figures. Demographic changes, such as population increase, have also appeared in districts located on the outskirts of the city. However, there is also the contradiction in recent years of high crime ratios being found on the outskirts of the city, especially in squatter settlements (Erkut, Ocakci and Unlu, 2001; Unlu, Alkiser and Edgu, 2000).

The Case Study Area

The case study area, 'Atasehir', is a new satellite-type settlement situated on the Asian side of Istanbul. Although planned for 18,000 apartment flats and 80,000 residents, only the eastern area of the settlement has been constructed, and that recently. Currently, 8,596 flats have been constructed, housing 35,000 residents. (Figure 4.1)

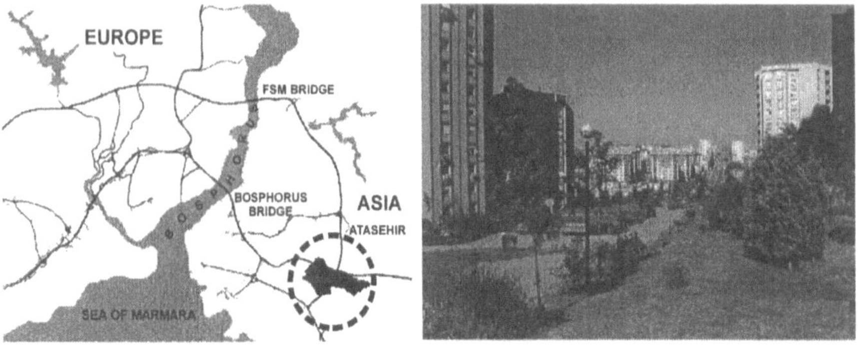

Figure 4.1 The location and the view of Atasehir

The settlement is mainly occupied by members of the upper-middle income group, thus the general design qualities are in accordance with these types of users' expectations. However, while the settlement creates the image of a planned housing area around a boulevard – a main axis – the area around the settlement is made up of unplanned and unauthorized houses, small manufacturing functions, storage facilities and office buildings. Therefore the life style in the settlement is at odds with the surrounding neighborhood. The non-existence of transition areas between the settlement and its surroundings causes feelings of insecurity, especially in the entrance and exit areas of the settlement.

Figure 4.2 The bumper zones of the settlement

The insufficiency of privacy mechanisms in bumper zones and the non-existence of transitional spaces inhibits residents' activities; rarely used public spaces between apartment buildings can be perceived and evaluated as potential crime areas. Unfortunately, our observations confirm the fact that a harmonious physical and social relationship between the settlement and the surrounding neighborhoods does not exist, and no planning strategy to overcome this problem has thus far been presented.

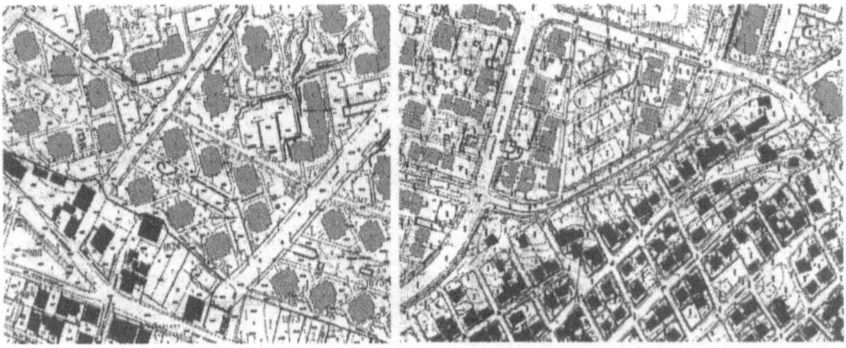

Figure 4.3 Intersection of two patterns

The insufficiency of planning strategies between two different urban environs was responsible for the growth of uncontrolled areas and settings, especially in the outer zones of the settlement.

The Method

The case study data was analyzed in three steps, based on crime frequency, space-syntax analyses and interviews with residents. The crime data was collected from the local police station and based on recorded crime frequencies between March 2001 and March 2002. The crime frequencies were categorized according to categories such as burglary, car theft, theft of items from within cars, and mugging. By indicating the rate and location of crimes within the settlement using the data provided by the police, a 'crime map' of the settlement was obtained. The residential zones of the settlement were consequently categorized as 'inner' and 'outer' zones.

In the second step, the variables analyzed also covered physical parameters like the spatial integration of residential zones, the syntactic integration or depth values of residential zones and the circulation routes of the settlements. The integration of zones was evaluated with the help of the space syntax approach. The syntax data was obtained from the 'Spatialist' software developed by Georgia Tech University. The program allowed us to ascertain the integration or depth values of the circulation axes of the site. This data is also very important in the understanding of isolated resident zones or more active zones based on physical variables. The determination of integration/depth values required the assumption of six intervals. The integration/depth values of the routes which linked housing units in the sample are determined in accordance with these intervals. The obtained data also gave us an idea about the integration or depth values of the sampled building within its immediate surroundings.

The third data group presented the users' evaluations about feelings of security. This data was collected through interviews with 80 occupants who were randomly selected from the inner and outer zones. The daytime and nighttime evaluations were separately indicated on questionnaires and five levels of security were determined on a semantic scale.

Consequently, the results of this study are categorized in three formats: residential zones based on crime frequencies, physical outcomes derived from spatial integration/depth values, and residents' evaluations on the feeling of security. All research data also tested the correlation between housing zones and occupants' feeling about security.

The Analysis of Determinants and Results

The data collected in three steps can be argued to be determinants vis-à-vis feelings of security. First of all, a variety of crimes were found to occur in the residential zones. The main crime types can be evaluated as burglary (theft from the house), theft from the workplace, car theft and mugging. These crime categories present unique characteristics due to the location, occurrence and post-event situations. However, this study concentrates on specific crime types rather than on whole categories of crime. For example, mugging always occurs in more populated and accessible areas; the escape route for the mugger is located around shopping malls and banks, and continues on main axes. This research concentrated on the outcomes of 'residential crimes' and eliminated some crime categories like mugging and theft from the workplace. Other crime types, like theft from the house (burglary) and theft from cars, were considered as part of the 'residential type of crime' in this study. The crime data which was collected from the police station points to the fact that residential-type crimes occurred mainly in outer isolated residential zones (Figure 4.4).

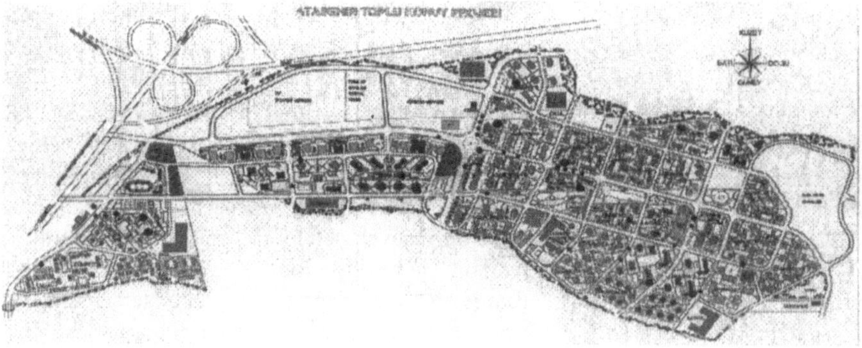

Figure 4.4 Distribution of crime frequencies

This study proposes a residential difference based on crime frequencies as indicated in Figure 4.4. According to the distribution of residential zone, the case study area is evaluated as inner zones and outer zones (Figure 4.5). As result of accessibility, factors like trading and shopping activities and population density in inner zones may cause a high public crime frequency but lower residential crime rates. In other words, while inner zones present lower residential crime frequencies, they also present higher crime ratios in public areas. This emphasizes the point that residential crime in outer zones is as high as assumed in Figure 4.5.

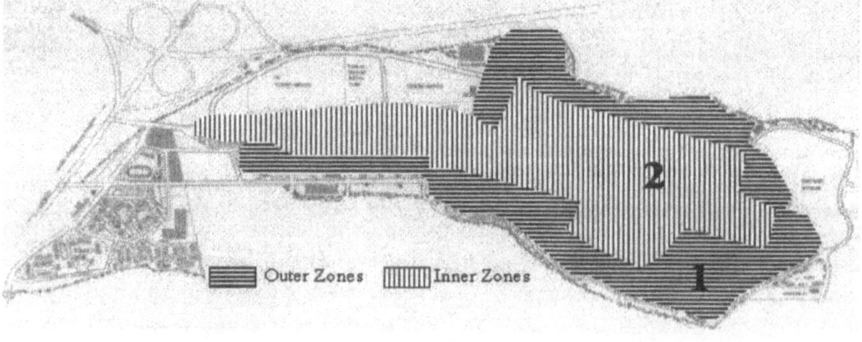

Figure 4.5 An assumption about the residential zones based on crime frequencies

When we analyze the crime rates in residential zones, there is a difference between the outer residential zones and the inner residential zones. While the residential crime frequency in inner zones is 40% of the total crime rate, this ratio encompasses 60% of total crime frequency in outer zones. The crime types also vary in accordance with residential zones. While car theft in inner zones accounts for 57% of the whole, this ratio is 43% in outer zones. Residential theft in the inner zone constitutes 25% of whole, this ratio therefore entailing 75% in outer zones. These figures also support the assumption indicated above, that the residential-type crimes are high in outer zones, 75% vs. 25%, while, conversely, car theft is high in inner zones, 57% vs. 43%.

The second data collection step of the case study is the examination of the physical pattern of the site vis-à-vis integration or depth values of residential zones. The syntax data, derived from the 'Spatialist' software of Georgia Tech University, USA, allows us to evaluate the integration or depth values of the circulation axes of the site (Figure 4.6). This data is also very indicative in understanding isolated resident zones or more active patterns based on syntactic variables. The established integration/depth values can be grouped in six intervals.

The integration values indicate physically more integrative circulation routes, and their values present high figures, while the isolated circulation routes lower figures respectively. This assumption is also relevant for depth values. Higher depth values give us the idea of isolated zones; lower values, however, indicate 'shallow' circulation routes, more active zones and intersected axes. The circulation routes are also indicative in understanding the nature of the physical pattern, and this variable can be considered a physical indicator for public and residential occurrences of crime. The space syntax analysis emphasizes the point that high integration values are found on the main axes, and they correspond to the first category. This category also covers the more integrated residential zones; conversely, the integration values of secondary routes decrease from the intersecting point of the main axes. Categories like 2 and 3 mainly define the branch routes, and their linkage with its vicinity do not present higher integration values, these corresponding instead to the levels of categories 4 and 5. The categories indicating lower integration values are mainly situated close to the boundaries of the settlement. As a consequence of low surveillance and low density, these residential zones can be described as isolated housing areas (Figure 4.6).

The interview stage, designed to determine feelings of security in occupants, was the third stage, the data collection stage of the research. This data was collected through interviews with 80 occupants who were randomly selected in inner and outer zones. In interviews, the views of the occupants regarding security at home, in the immediate surroundings of the home, the settlement and the immediate surrounding of the settlement during the day and night were requested (Table 4.1).

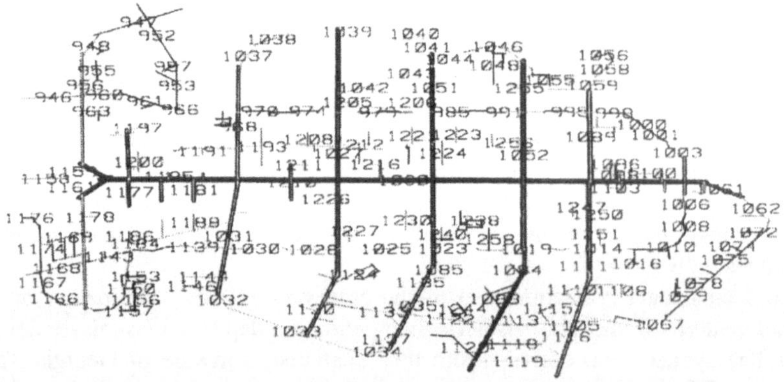

Figure 4.6 Spatial integration and depth analysis of the settlement

Table 4.1 The occupant's evaluations about the house and close surroundings

		DAY						NIGHT					
		Very secure	secure	Less secure	insecure	very insecure	Σ	Very secure	secure	Less secure	insecure	very insecure	Σ
House		45	26	7	2	0	80	40	34	4	2	0	80
	%	56	32,5	9	2,5	0	100	50	42,5	5	2,5	0	100
Around the House		29	25	20	4	2	80	22	28	23	5	2	80
	%	36,5	31	25	5	2,5	100	27,5	35	28,5	6,5	2,5	100
In Compound		26	26	21	5	2	80	14	28	26	10	2	80
	%	32,5	32,5	26	6,5	2,5	100	17,5	35	32,5	12,5	2,5	100
Around the Compound		0	0	10	37	33	80	0	0	7	24	49	80
	%	0	0	12,5	46,5	41	100	0	0	9	30	61	100

In light of the data analysis, 'the feeling of security' in residential zones can be argued as consisting of an interaction of three determinants:

- the 'residential zones' based on crime occurrences,
- 'the integration or depth values of axes' linking the residential zones,
- psychological and sociological aspects of 'feelings of security' based on the temporal order.

In this theoretical framework, significant chi-square correlations due to these determinants are revealed. There is a high correlation between occurrences of crime and the feeling of security in accordance with the daytime and nighttime evaluations. For instance, occupants living in outer zones declared that the feeling of insecurity at night was comparatively high, and the correlation between residential zones and security at nighttime is statistically significant ($x2= 15,362>11,070$, df=5, $p=0,009<0,01$). This judgment is not relevant with regards to daytime differences. While 74 % of residents who live in inner zones sensed a secure atmosphere during the day, this ratio is in a descending trend, with 50 % of residents in outer zones registering complaints about daytime security. These values also show a significant relationship between residential zone differences and the occurrence of crime during the daytime ($x2=11,823>11,070$, df=5, $p=0,37 <0,05$).

There is another chi-square correlation between spatial integration values and the existence of feelings of security. The evaluations reveal a high feeling of security, especially during the daytime, among residents who live in more integrated areas, such as in inner residential zones ($x^2=35,877>31,410$, df=20, $p=0,016<0,03$). Residents who live in more segregated residential zones present lower figures in terms of feelings of security. The data in relation to the nighttime reveals contradictions among the residents. While the residents who live in inner zones stated they sensed a more secure atmosphere during the night, residents who live in isolated regions and routes conversely declared lower values of security ($x2=34,157>31,410$, df=20, $p=0,025<0,03$).

On the other hand, this study illustrates a significant relationship between the residential zones and the integration/depth values of circulation routes. The chi-square correlation between these two residential zones and syntactic parameters is highly significant ($x2=50,311>9,488$, df=4, $p=0,00< 0,01$), implying the crime rate is very high in outer zones or in more segregated routes, while having lower values in inner zones.

The settlement in this case study presents an interactive model of security that comprises the chi-square correlations between factors such as the residential zone, integration/depth values and residents' evaluation of crime (Figure 4.7).

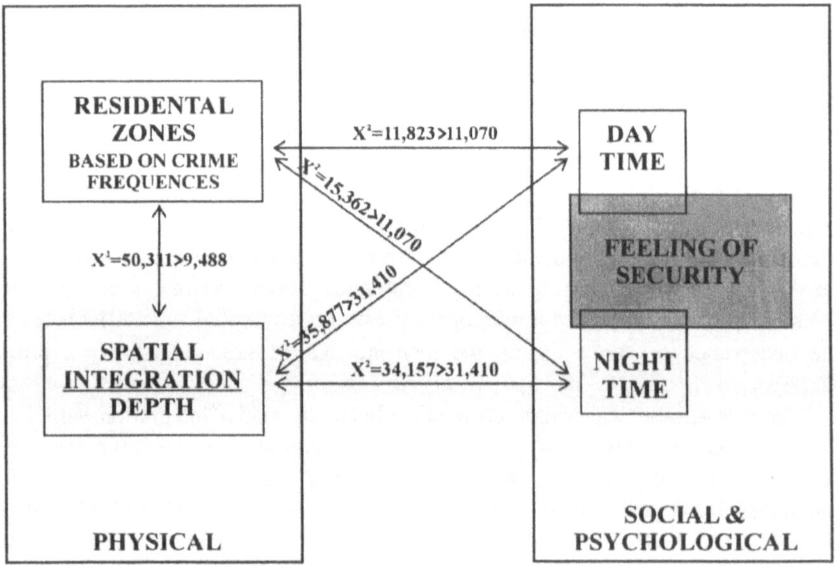

Figure 4.7 The conceptual model on 'feeling of security'

This interactive model presents the notion of crime as being not merely a socio-psychological issue but as one whose determinants are rooted in the physical milieu, like residential zoning, integration and the depth values of space syntax. This study once more emphasizes crime vulnerability in segregated residential areas and thus contains echoes of Hillier (1998) and Shu (1999). In recent analyses, they argue that segregation (between neighbors) and cul-de-sac patterns are quite vulnerable against crime; our paper, however, topologically, culturally and spatially indicates a quite different urban pattern. This research, like others, once more emphasizes the lack of surveillance and social relations in more segregated residential zones. Divergent concepts, however, such as more integrated urban patterns, may create much more socially active places, with neighborhood control and surveillance.

Conclusion

This chapter proposes the importance of the evaluation of crime profiles in physical settings and, specifically, feelings of security within this concept. The realization of secure home environs in which the social and physical integration processes interact may be put forward as resolutions to this conflict. Of course, these approaches lead us to conclude that macro-planning processes are necessary in the planning and design of residential environments. Integrative planning strategies force us to reconsider the establishment of social spaces, especially in transition areas. The physical staging in transition areas and the creation of sociopedal environs can be primary solutions by consciously integrating social and physical parameters. In this way, new planning strategies will overcome emerging social problems, especially in the transition areas, and the spatial solutions will minimize the physical and social distinctions between different home environments, eventually and hopefully leading to a decrease in residential crime.

References

Apak S. (1998). *A Conceptual Model for Obtaining Secure Environments in Mass Housing Areas*, Institute of Science and Technology, Istanbul Technical University, Doctoral Thesis, (in Turkish).

Centel, N. (2000). 'The Reason and Criminological Explanation of Increase in Property Crime', *The Seminar on Property Crime Prevention*, Department of Security Publication. Republic of Turkey, Istanbul: Adım Ajans, Publication Number: 8, pp. 56-67 (in Turkish).

Erkut, G., Ocakci, M. and Unlu, A. (2001). Evaluation of Crime Profile in Istanbul Metropolitan Area, *Trialog* 70, pp. 30-33.

Hillier, B. (1988). Against Enclosure. In N. Teymur, T. Markus, and T. Wooley (eds.), *Rehumanizing Housing*, (Vol. 2, pp. 25.1-25.11). London: Butterworths.

Lawson T. and Heaton, T. (1999). *Crime and Deviance*, London: Macmillan Press Ltd.

Laymon, R. S. (1974). Architectural Design and Crime Prevention, *EDRA 5*, Environmental Design Research Association Inc., vol. 3, Human Factors, pp. 43-50.

Marcus, C. C. and Sarkissian, W. (1986). *Housing as If People Mattered. Site Design Guidelines for Medium-Density Family Housing*, Berkeley and Los Angeles, California: University of California Press.

Newman, O. (1973). *Defensible Space, People and Design in the Violent City*. London: Architectural Press.

Shu, S. C. (1999). Housing Layouts and Crime Vulnerability, *Space Syntax Second International Symposium*, Brasilia, Brazil, vol. 2, pp. 25.1-25.12.

Taylor, R.B. (1987). Toward and Environmental Psychology of Disorder: Delinquency, Crime and Fear of Crime. In D. Stokols and I. Altman (eds.), *Handbook of Environmental Psychology*, (Vol. 2, pp 951-986). New York: John Wiley and Sons Inc.

Taylor, R.B. (1988). *Human Territorial Functioning*. New York: Cambridge University Press.

Unlu, A., Alkiser, Y. and Edgu, E. (2000). *An Evaluation of Crime Patterns in the Context of Physical and Socio-cultural Change in District of Beyogl.*, Research Project Report No:1094, ITU Research Fund, Istanbul.

Unlu A. and Edgu, E. (2001). Residential Areas in City Centres and Crime, *Istanbul*, 38, 86-88 (in Turkish).

Chapter 5

Transfer Process of Self-Built Houses in Environmental Protection Areas in the Region of Campinas, Brazil

Silvia A. Mikami G. Pina, Doris C.C.K. Kowaltowski, Regina C. Ruschel, Lucila C. Labaki, Stelamaris R. Bertolli, Francisco Borges Filho and Édison Fávero

Introduction

In developing countries and particularly in Brazil, low income population groups often occupy vacant land close to urban areas, with good transportation infrastructure and job opportunities. These areas are often under environmental protection and therefore residential building activities are illegal. In many municipalities these so-called urban invasions fall under the scrutiny of environmental laws and the population is forced to move to new locations through government supervised programs.

This chapter describes the transfer process of one of these groups of populations in the region of the city of Campinas, in the State of São Paulo, Brazil, specifically the families of the *'Jardim Conceição'*. The local government provided a new area for subdivision and organized the distribution of residential lots per families. The program further offered the option of house constructions through a minimal house 'Kit', called 'COHAB-Embrio'. The 'Kit' supplied building materials on-site. Families also had a choice of using the assistance of a technical aid program, devised by a research group of the State University of Campinas (UNICAMP), called 'TITAM'. Houses were to be self-built in both cases.

The 'COHAB-Embrio' consists of a minimal construction of 28m^2, with a small multipurpose room, kitchen and bathroom as shown in Figure 5.1. The quality of the construction is precarious especially since no ceiling is included, an important thermal comfort factor for the local climate. The choice of building materials such as concrete block and thin asbestos cement roof sheeting also cause thermal comfort problems.

Figure 5.1 Floor plan of the COHAB 'Kit' house

The technical support program, called 'TITAM' (*Transferência de Inovação Tecnológica na Autoconstrução de Moradias - Transfer of Technological Innovation in Self-Built Housing*), consists of providing self-builders with a house plan, based on the lot conditions and family needs and wants. In the particular case of the assistance given to the population of the '*Jardim Conceição*', two house types were preferred or selected as most appropriate for the lot conditions and family needs as seen in Figure 5.2. The house design types were based on an extensive research project in relation to housing needs of this type of population in the Campinas region (Kowaltowski, 1998 and Kowaltowski et al., 1998). To be able to assist a large number of families, design projects are produced by an automated, CAD based, house design program called 'AUTOMET', especially created for this purpose. The research methods applied to these investigations were described previously, especially in Kowaltowski, et al., 1995 a, b and Kowaltowski, et al., 2000.

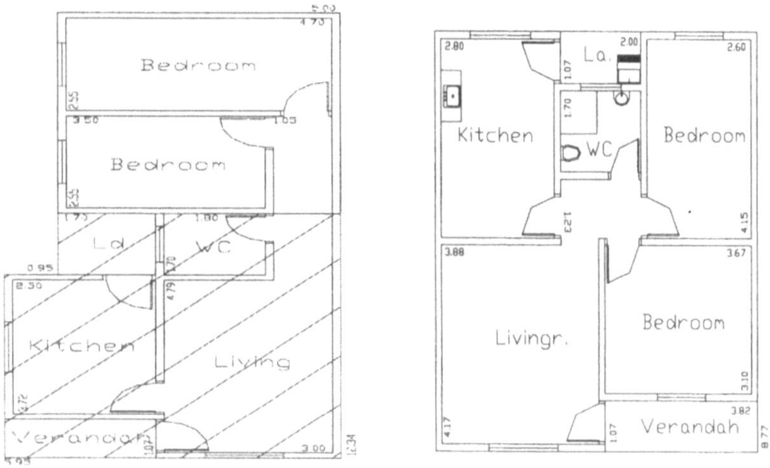

A B

Figure 5.2 Two preferred house floor plans of the 'AUTOMET' design program used in the 'TITAM' program of the *Jardim Conceição* example[1]

The experience of cooperating with a low income population in a situation of fast action for the relocation of dwellings is evaluated in this paper. Previous investigations, accumulating data on the self-built house process in Brazil, show that the self-built housing production has specific characteristics and problems.

Background to the Self-Building Process

In Brazil, distinctions can be made between the housing process of low income groups and middle and wealthy social classes. Due to specific local economic and social structures, as well as urban growth patterns, self-built houses, i.e. houses built by owner families, make up a substantial percentage of Brazilian housing production (Nolasco, 1995; Kowaltowski et al., 1995 (a), Ornstein et al., 1995, Werna, 1996). Self-built houses are the predominant mode of urban habitation production of the low income population in other Latin America countries as well (Turner, 1976 and Kellet and Napier, 1994). According to Brazilian data around 60% of the local housing production is self-built (Schulz, 1996).

1 The hatched area of the Plan A indicates the minimal construction, which some families built first. Plan B is the preferred house design in the Campinas region for the self-building population. Drawings are not to specific scale.

The self-help activity has been hailed by many authors as a correct approach towards solving enormous housing deficits. On the other hand, the acquisition of finished houses is preferred by the public (Hamdi, 1991). A very low income population, however, cannot afford them. The monthly income level of urban self-builders in Brazil lies between 3 to 10 times the minimum salary, the minimum salary (SM) being equivalent to around US$ 75.00. A study on housing characteristics in Brazil divides low-income levels into 3 groups. Thus a very low income corresponds to a salary from 0 to 2 SM, low from 2 to 3 SM and medium low from 3 to 6 SM (Brusky and Fortuna, 2002).

In Brazil, 30 times more homes are being built in the informal as against the formal ways of construction (Augusto and Bastos, 1997). A distinction must be made between self-building activities on land without tenure such as invasions and slums or the so called 'favelas' and those on lots acquired by families in private subdivisions or through government distribution programs. As in many developing countries, spontaneous housing, without tenure, is synonymous of extreme conditions of poor quality housing and has a negative impact on the urban environment (Pettang and Tatietse, 1998).

For the self-building process, with tenure on acquired land, the income range of families extends to 10 times the minimum salary per year in some areas. A recent study in five government sponsored subdivisions in the city of Campinas, in the State of São Paulo, demonstrates that 12% of self-builders declared incomes above 5 minimum salaries (SM), 32% stated to be in the range of 3 to 5 SM and 48% in a lower range from below 1 to 5 SM. (Watrin, 2003). Even though the income range of self-builders is wide, these earnings are insufficient for the acquisition of a home through the regular housing market, aimed at middle and upper income level classes.

Local socio-cultural factors also play a part in the preference shown by families in building their own homes. No national housing program has been adopted in Brazil since 1986, and this has led to an enormous housing deficit (Valença, 1992 and 1999). During the last forty years as well, towns like Campinas, a city of around one million inhabitants about 100 kilometers from São Paulo, the largest city today in Brazil, have doubled in size in a decade (Patarra et al., 1994). This growth occurs mainly at the fringes of the city, which are being occupied by low income populations. Due to the speed of urban growth, land use planning encounters difficulties in controlling land occupation and construction quality. Access to urban land spurs speedy construction of minimum houses by self-builders. This building process lacks a proper design and planning stage and results in many transformations of houses during the lengthy construction period (Kowaltowski and Pina, 1995 and Tipple, 2000). However, most self-built settlements can be described as fairly organized suburbs, after consolidation (Kowaltowski et al., 1995 (a)).

The self-building process has been described as belonging to vernacular architecture or what is termed new traditional environments, but this new tradition must be qualified (Rapoport, 1988 and Kellet and Napier 1994). Studies have shown that vernacular architecture in many places is based on profound elements which embody environmental quality. The self-building process in Brazil, however, has specific characteristics and problems. Due mainly to low quality design solutions self-built houses present on the whole a low environmental comfort standard (Labaki and Kowaltowski, 1998). The local new vernacular thus lacks some of the positive elements of many traditional buildings, especially praised for their intelligent solutions to climatic problems.

In this paper we discuss house production in legal situations through the self-building process of individual families and their need for technical support.

Assistance Method Used in the 'Titam' Program

In Brazil many programs exist which offer help to low-income families in their home acquisition process. Some programs offer financial help, others distribute urbanized lots for individual self-building activities. Many programs are aimed at slum eradication through relocation of populations, especially of slums that occupy risk areas. House construction is often organized through a cooperative, or participative, building process, known as 'Mutirão' in Brazil (Cabannes, 1997). Other programs aim to improve spontaneous settlements through the provision of urban infrastructure. NGO's have been involved in housing assistance programs throughout Brazil. Most Brazilian city administrations provide programs which distribute standard house plans to families. In Campinas the local administration extended this program through a partnership called PROMORE, to include exemption of fees and access to professionals of the syndicate of civil engineers for families earning less than 5 minimum salaries (Campinas, 2002). Few self-builders use the standard house plans, for fear of increased taxation on their properties and a general lack of incentives to legalize the construction of houses.

The long term investigations of the self-building process in the region of the city of Campinas, carried out by the team of this report, led to a characterization of the local self-building process. Aware of problems with the self-building process, especially in the Campinas region, a team of researchers from the State University of Campinas (UNICAMP) set out to devise a design assistance program. The team was composed of architects, civil engineers and researchers in environmental comfort. This team developed an automated house design program 'AUTOMET', as well as the organization of a cooperative technical aid program.

Groups composed of students and researchers advised families, on-site, on their house design needs in the *Jardim Conceição* relocation program situated in the district of Souzas, Campinas. Examples of the on-site assistance meetings are shown in Figure 5.3. A specially adapted car, donated by General Motors of Brazil for this purpose, was used in this program. Appropriate equipment was made available on-site, such as a table with a connected computer and printer. Due to these assistance conditions, designs were able to be modified to specific needs and wants on-site during these visits. Self-builders were provided with a simplified house plan and a perspective drawing with specific advice on the roof configuration and window locations.

A B

Figure 5.3 'TITAM' – A: On-site advice for low income families on house design needs and B: a preliminary meeting scene

Through this aid program 84 families were attended directly, on-site, out of the potential 120 families of the new subdivision. Assistance was made available on Saturday mornings over a two-month period. This period was found to be the most productive in finding families at home and able to participate in the program. The assistance period extended over two months due to the fact that in some cases designs were individualized for some families, which made the scheduling of several meetings necessary.

House Evaluations

During the first six months of building activities in the new location (*Jardim Conceição*) frequent visits were made by the 'TITAM' team to assess the construction process and progress. A questionnaire to evaluate the construction phase was applied to all families present. This questionnaire registered the origin of the families, their socio-economic level, the house design planning that took place prior to construction, construction experience of each family and the type of help sought in order to proceed with the construction of the house.

User satisfaction was assessed with regard to the design of the house and its construction. Houses were evaluated as to construction techniques employed, stage of house construction and with regard to changes introduced to the designs during the construction phase. All the changes introduced were registered in CAD designs and analyzed as to their purpose. Ten houses were specifically chosen for a more detailed evaluation. In these cases, environmental comfort conditions were measured with technically appropriate methods and equipment. Due to problems of access, one of the houses was finally excluded from this in-depth evaluation.

Results showed that after six months of relocation of the specific community, 40% of the lots were still empty with no signs of building activity. Of the families who had relocated, 45% opted for the 'COHAB-Embryo' project. 25% of families designed their own houses without reference to either the 'COHAB' or the 'TITAM' models presented to them. 30% of the families assisted in the 'TITAM' program opted for the design individually produced for them by the UNICAMP team. The houses not based on either the 'COHAB' or the 'TITAM' assistance plan have not as yet been technically evaluated or their satisfaction measured. The primary purpose at this stage was to measure the assistance programs as such, the relation between design assistance and housing quality.

The 'TITAM' houses were built by families themselves or with some hired help. Satisfaction was assessed through various methods. Personal responses in a post-occupancy evaluation were registered. An analysis of modifications introduced to the design during the construction phase was undertaken and technical environmental measurements, detailed later in this paper, were evaluated with respect to recommended comfort standards.

Of the 'COHAB' houses, 58% were modified in the first six-month period of construction. These modifications consisted in precarious additions, such as an extended roof under which to locate laundry activities. The 'TITAM' constructions were true to design outlines in 72% of cases. Alterations were introduced in some cases due to inadequate perception of room sizes during the construction phase, which can give the impression of reduced functional space. Thus, some families eliminated the front verandah, present in all 'UNICAMP' designs for privacy and comfort purposes, incorporating its area into the interior of the house, especially the kitchen. Some families also judged the bathroom to be small, although it was larger in area than the bathroom in the 'COHAB' houses being built by some of the neighbors. These changes are shown in Figure 5.4. With the incorporation of the small external covered verandahs at the main entrance and at the kitchen back door, houses lost important transition spaces.

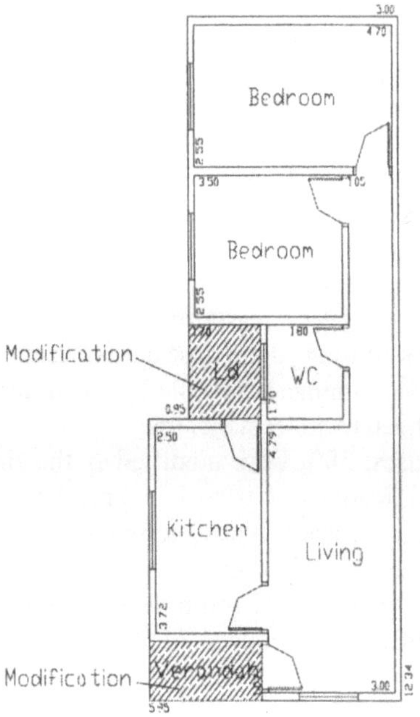

Figure 5.4 Example of floor plan changes introduced to the 'TITAM' houses in the *Jardim Conceição* Assistance program

These modifications, according to the assistance team's technical analysis, reduced the quality of functional aspects of both the living room and the kitchen. Circulation patterns were affected and privacy was reduced. The front and back door, as well, lost their important rain protection element. The living room window, in most site conditions, was no longer protected against excessive sun exposure, a major thermal comfort factor in the local climate. These results caused the 'TITAM' team to reflect on the need for further information to be added to both the drawings of the houses and technical assistance material distribution in the form of booklets, described further on in this paper. Drawings were enriched with better furniture arrangements. The information manual discusses, among other things, aspects of privacy, functionality and technical elements of rain and sun protection.

Further analysis of the houses showed that 61% of self-builders used concrete block for the construction of walls. The majority of 'UNICAMP' houses, however, were built with ceramic material, a local building material with better thermal resistance. In relation to ceilings, important in the local climate as an element to reduce heat gain through the roof, only 30% of constructions introduced this protection through wooden or prefabricated concrete ceilings. Here again, one of the reasons for the low occurrence of this important comfort element was related to the 'COHAB' kit, which does not include a ceiling. But the fact also points to the need for more information on the relation between comfort and construction techniques among this type of population. The first priority of a roof is seen as providing shelter, and not the thermal comfort that a house can provide when finished and occupied. Furthermore, one must reflect on the fact that the majority of the families in the specific population had low comfort expectation, coming as they did from extreme conditions of huts built of thin wooden walls and low asbestos sheet roofs.

To determine the sample of ten houses to be examined specifically for environmental comfort conditions through technical measurements and observations, a three point method was devised for selection:

- A fairly equal representation of 'COHAB' and 'TITAM' constructions;
- Similar siting conditions in the same street to represent equal solar orientation situations;
- The streets chosen should have similar ventilation conditions and relation to noise sources, to assure equal acoustic and thermal parameters.

Figure 5.5 shows the map of the subdivision and an indication of houses evaluated. Accordingly, the street named *Jõao Maria Batista* is the main access to the area, where car and pedestrian traffic can be observed. On the street named *Avenida 2*, houses occupy only one side of the street and for the street called *Particular 2* the research team observed unfavorable ventilation conditions.

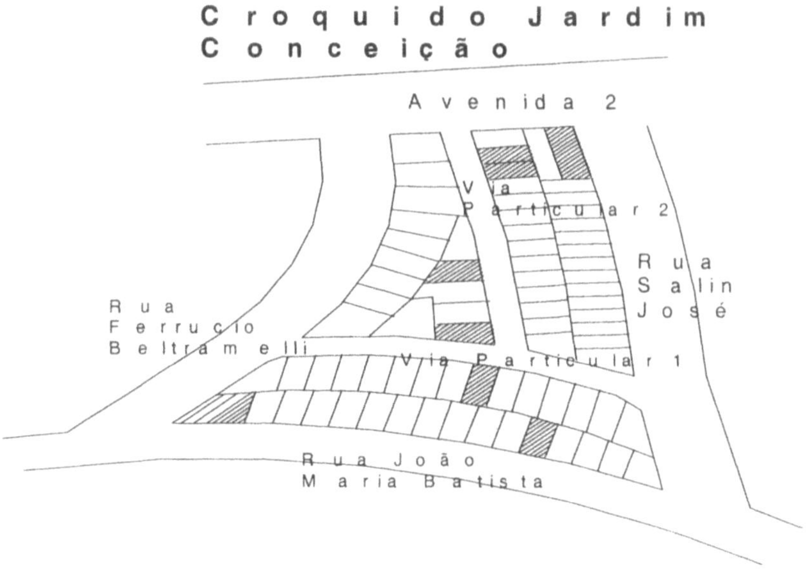

Figure 5.5 Map of the subdivision called *Jardim Conceição*

Various methodological problems had to be overcome so that some comparisons could be made. The room sizes are not the same in the two types of constructions evaluated. The 'COHAB' kit houses were evaluated in the main room and the kitchen and in special cases in an added bedroom. The 'TITAM' houses were all evaluated in the living-room, kitchen and one bedroom. The thermal parameters evaluated were the globe or radiant, dry and wet bulb temperatures and the air speed. Equipment used for these measurements were: a digital thermometer, (TGD-200, Instrutherm) and an anemometer (AM-4204, Lutron). Three locations of the equipment were adopted in each case, the center of the room, close to the window and in proximity to the back wall. The satisfaction evaluation was based on the 'Fanger' Estimated Vote Method and used light clothing and relaxing as the activity index (Fanger, 1972). The Fanger Method was used to obtain an evaluation of the real comfort conditions, since the first satisfaction enquiry yielded only positive responses by self-builders. The positive satisfaction is seen as a normal reaction in the situation in which these families found themselves, having just gained their own homes through relocation and the self-building process.

To assess acoustic conditions in the houses, the noise level was registered through sound pressure measurements (NPS) using an integrated pressure reader (00026, Robotron) with an 'A' scale compensatory setting. Measurement procedures followed the Brazilian acoustic code (ABNT, 1987). The equipment was placed at a 1m distance from walls and a 1.5m distance from windows, in both cases at a height of 1.2m. Satisfaction ratings were based on comparisons with recommendations found in the Brazilian code.

Visual comfort conditions were measured through the light level (I), the quantity of lux in the rooms, using a luximeter (LX-102, Lutron). Measurements were taken at 5 points: the center of each room and four points around the center at a 50cm distance. Satisfaction ratings were based on comparisons with recommendations found in the Brazilian Sanitation code.

In relation to summer thermal comfort conditions eight houses have a higher than 50% dissatisfaction estimated vote. Only one house, built according to the 'TITAM' design, as shown by the full plan configuration in Figure 5.2 A, was considered reasonably comfortable in summer climatic conditions. This particular house was one of the few constructions with a ceiling and built with ceramic block walls.

Further observations and technical data must be added when analyzing thermal summer conditions and assessing the problems encountered. In many instances the team encountered houses with closed windows and in some cases large pieces of furniture, such as closets, were positioned causing window openings to be blocked. Most houses had no exterior finishing and external colors thus were dark gray or brick-color, poor conditions for heat reflection. The choices made by the population in relation to roofing material and window sizes were also shown not to be ideal.

In relation to acoustic conditions results showed noise reverberation to be satisfactory in all but one of the 'COHAB' kit type houses. In the 'TITAM' houses the kitchens showed higher reverberation rates and thus unsatisfactory conditions. A detailed analysis of this data showed that the 'COHAB' houses do not have specific separated kitchens and rooms, since there is only an opening between the two spaces. The total house space is extremely small, in relation to family size. The rooms were very densely occupied with beds, clothing and other sound absorbing materials, like sofas in the kitchen, and mattresses on the floors. The 'TITAM' houses, larger in size, were sparsely furnished and clothing was kept in bedrooms, where acoustic conditions were not compared.

In relation to lighting conditions all but two of the 'COHAB' kit houses could be considered satisfactory. The two inadequate cases were those where a window was blocked by furniture. Since the 'COHAB' kit included internal paint on the walls, the light reflection conditions are superior in relation to the 'TITAM' houses, none of which at the time of analysis had internal finishing and therefore dark colored walls.

Owner expressed opinions showed an overall positive satisfaction with the house of either type. When specifically asked about comfort conditions some conflicting results appeared. Thus families with larger houses showed dissatisfaction with the room sizes and families in overcrowded houses indicated no problems with the area of the house. Many owners also were reluctant to express their opinions.

Discussion

Results indicated that many factors influence satisfaction levels. The community spirit at the time of data collection must be taken into account. Our investigation showed that some dissatisfaction was due to problems with urban infrastructure in the subdivision. Roads were not paved, the promised sewer system was not installed and water and electricity connections were still unsatisfactory. These conditions cause friction between the community leadership and individual families and influence the expressed satisfaction of owners.

Final results showed a greater satisfaction with the spaces provided by the 'TITAM' designs. Post-occupancy evaluations and environmental comfort measurements indicate that these houses provide a better quality of life for low income families, based on adequate functional space provision. Thermal, acoustic and lighting conditions are shown to be related to a wider range of aspects. These include technical choices and user habits. Proper orientation and dimensions of openings indicated on designs are insufficient elements to ensure comfort. Choices of construction materials and techniques are important factors to provide comfort quality, as are the uses and occupation types of spaces. Thus if windows are blocked or never opened, for reasons of problems with security or dust, thermal conditions will be negatively affected, independent of design orientation. From these results the assistance team concluded that more information is needed to ensure a higher level of house construction quality.

The specific assistance experience also caused the team to make some procedural changes in the technical aid program of 'TITAM'. The presentation of designs included typical local furniture arrangements in each room. The inclusion of furniture was already part of the 'AUTOMET' solutions, but the furniture arrangement shown was considered, after this experience, to be insufficient to transmit the correct perception of space in the future house. This change in presentation should avoid some of the design modifications introduced by builders during the construction phase.

A preliminary explanatory visit to a potential community was considered important. This visit should include a full explanation of the assistance program. The house design choices are to be presented and their functional and environmental comfort goals explained. Questions should be answered and reduced scale three dimensional models should be presented (Figure 5.6). These models can better illustrate the type of house that will be able to be built according to the design instructions which are distributed. In an assistance program for another community, albeit not in the same condition of relocation as described in the case of the *Jardim Conceição* population, the 'TITAM' team organized such a preliminary meeting as shown in Figure 5.3 B. The community, in this case, showed greater interest in the project and fewer needs for design modification were registered. Families had a better understanding of the designs of the houses through the three dimensional models and were made aware of aspects of environmental quality of their future homes.

Figure 5.6 3D models of house designs

Information, in various forms, must be made available to self-builders. Assistance programs must be on-site during a period of the construction phase to discuss choices of materials and building techniques. The technical reasons behind design features must be made clear in order for these to be preserved when construction takes place. These features must gain importance to avoid the elimination of positive elements and changes which may diminish certain comfort aspects. The elimination of verandahs is such a case. Verandahs are a traditional element in Brazilian colonial architecture (Labaki and Kowaltowski, 1998) and have positive effects on shading conditions of outer walls. Verandahs protect the front door from rain and increase privacy in small houses by providing a transition space.

Reducing the size of windows and changing their location, thus excluding the possibilities of cross ventilation, are other important factors to be discussed with self-builders. The importance of ceilings must be stressed and the proper choices of building materials must be indicated.

Three different ways of disseminating information were discussed and developed as illustrated in Figure 5.7. First, a booklet in the form of a design and construction manual was created. Second, a *WEB* page was considered important today, to widen the scope of interest in the quality of self-built houses in Brazil. Thirdly, an assistance system in the form of a 'construction clinic', where doubts regarding building technique can be raised, discussed and solved, should be made available in every city with large self-building activities.

Figure 5.7 Illustration of information material for self-builders

To determine the scope of the information material, a thorough analysis of the self-building process was made. Changes introduced by builders to designs were especially analyzed. This evaluation was considered important, and led to changes in some of the designs provided in assistance programs and made it possible to indicate to self-builders the positive features of each house design.

Secondly, important environmental comfort information was assembled and organized for communication to self-builders. Here some technically doubtful beliefs, such as the positive effect of room height on thermal comfort, were collected and analyzed as to their true value. In the case of the height of rooms, research shows that, when roof and ceiling construction conditions are adequate, the room height has little effect on thermal conditions and thus is not an important factor in relation to house construction quality. Of course minimum room heights must be observed for reasons of correct volume configuration, functional and ergonomic conditions.

The presentation form of the information material was researched to stimulate the use of such material and communicate adequately the content to the population in mind. Thus scope, language and illustration types were carefully studied. Erroneous interpretations were tested to ensure success.

In the booklet, or manual, the main purpose was to transmit to self-builders the importance of using a good design as a basis for construction. The elements of the 'AUTOMET' designs were therefore discussed one by one. Lot conditions, siting, orientation of the house and the functions and comfort conditions of each room were discussed. Room sizes are shown to be important for the proper performance of activities and arrangement of furniture. Thus the kitchen needs a certain size for cooking and eating to avoid the unnecessary incorporation of the verandah. Particular attention is given to excess sun exposure of windows and the necessity for shading through roof overhangs, verandahs and trees. Treatment of open or garden areas is also shown to influence directly the thermal and acoustic conditions of a home. Some hints are given on circulation and privacy in the home and shown to be related to the position of doors and the distribution of spaces in a house, as well as the existence of transition spaces, which thus should not be eliminated or incorporated into other spaces.

To make the distribution of such a booklet possible in relation to cost, the size was reduced and only three colors were used. The final illustrations are being produced by a member of the team at this time to ensure artistic value of the material. Distribution of the manual is seen as important in future technical aid programs in conjunction with on-site design assistance. Booklets should also be freely available in construction material shops.

The team also developed a specific environmental comfort booklet for children, as presented in Kowaltowski, et al., (2002). This booklet is an important contribution to increase interest in this subject at an early age. Many children in public schools come from self-builder families, thus this booklet was seen as a way of bringing this knowledge into the home, via the school. The form of information transfer was shown to be positive, when the school manual was tested.

Final Considerations

This paper describes a technical assistance program in one particular instance, organized by a team from UNICAMP. On the whole, this team has given such assistance, to this date, to around 400 families. Individual help is given to families who seek the University for information. Specific joint programs with housing agencies have been developed. In the particular case discussed here, the team worked in conjunction with the local city administration and a specific community.

From the experience of the case described above, many lessons have been learnt. First and foremost the importance of the distribution of house designs to self-builder was confirmed. In Figure 5.8 a house built with the assistance of this program is shown. Even though not yet finished, the result shows the potential for houses with expected comfort quality.

Figure 5.8 Example of 'TITAM' assisted self-built house

The goals of the assistance program were reached. Giving a segment of a low income population the opportunity for a better quality of life and avoiding waste in the construction process of self-built houses was the primary goal. The automated design method 'AUTOMET' was shown to be efficient in producing architectural designs with a degree of individualization and appropriate for site, climatic and economic conditions.

On-site presence was important for both the participant families and for the research team. Reality was present in all discussions and decisions. Interaction between self-builders and Architecture and Civil Engineering students created a link for further developments of technical information communication.

Other observations are important when evaluating a technical assistance program for the self-built house process in developing countries. This process usually starts through the acquisition of a building lot in a private subdivision or a subsidized government program. The legal status, achieved through the acquisition of a lot, creates expectations that may not always be able to be fully realized by these low income families. Lot dimensions are small, with a geometry that limits design possibilities. Lots are mainly rectangular narrow strips of land to reduce street front dimensions and thus subdivision infrastructure costs. Often the subdivision layout is not ideal for the siting of desirable house designs. Orientation of streets does not take into account sun exposure and ventilation conditions. Possible house design solutions are thus already reduced or flawed through the underlying subdivision design.

Self-builders also have expectations of solving more than their own housing problems. Thus most self-builders have 'dreams' of building several houses on the newly acquired lot. These additional constructions are desired to house other than immediate family members or to provide an extra income, through renting added space or the establishment of some commercial activities. In most cases a self-builder's main objective in building a home is not comfort, but to solve basic shelter and income problems. These factors must be taken into account when technical advice is given.

The individualization of designs is seen as important and was incorporated in the 'AUTOMET' design program. Individual dreams, however, are more diversified and not always technically possible or desirable from the standpoint of living standards. Individual design assistance, on-site, is thus more important in relation to providing technical information than in incorporating all individual desires into the new architectural form, which may not be ideal for environmental quality reasons.

The experience in the assistance program provided for the population of the '*Jardim Conceição*' has shown, as well, that on the urban scale different actors must play a role. First, the planning of subdivisions must be improved to provide a better basis for the quality of design of individual homes. Second, the basic infrastructure must be provided from the start to improve the urban image and living conditions. Lastly, a community spirit should be nourished through various methods. Provision should be made for open space for recreational activities. Subdivisions should be planned with the incorporation of community buildings, such as schools. Attempts must also be made, by local government or other organizations, to create community groups with special interests to improve the built environment.

References

ABNT, Associação Brasileira de Normas Técnicas (RJ) (1987), *Avaliação de ruído em áreas habitadas visando o conforto da comunidade*. NBR 10151. Rio de Janeiro: ABNT.

Augusto, C. and Bastos, R.(1997). Moradia Informal Vive Boom em São Paulo. In *O Estado de São Paulo (newspaper)*, Caderno Cidades, 03.08.97, São Paulo, pp. C1.

Brusky, B. and Fortuna, J. (2002). *Entendendo a Demanda para as Microfinanças no Brasil: um Estudo Qualitativo de Duas Cidades*. Rio de Janeiro: PDI/BNDES.

Cabannes, Y. (1997). From community development and mutirão to housing finance and casa melhor in Fortaleza, Brazil. *Environment and Urbanization*, 9(1), 31-58.

Campinas, Coordenadoria Setorial de documentação (2002), *Decreto No. 14 153*, 14 de Nov., Convênio, Município de Campinas e Sindicato dos Engenheiros do Estado de São Paulo, PROMORE, Programa de Moradia Econômica.

Fanger, P.O. (1972).Thermal Comfort, Analysis and Applications in Environmental Engineering. New York: McGraw-Hill.

Hamdi, N. (1991). Housing without Houses: Participation, Flexibility, Enablement. New York: Van Nostrand Reinhold.

Kellett, P. and Napier, M. (1994). Squatter Architecture as Vernacular: Examples from South America and South Africa. *Traditional Dwellings and Settlements Working Papers Series*, Vol. LX, 1-48, pp. 1-34.

Kowaltowski, D.C.C.K. and Pina, S.A.M.G. (1995). *Transformações de Casas Populares: Uma Avaliação*, Proceedings of the III Encontro Nacional e I Encontro Latino-Americano de Conforto no Ambiente Construído, Gramado, Brazil: ANTAC.

Kowaltowski, D.C.C.K. (1998). Aesthetics and Self-Built Houses: an Analysis of a Brazilian Setting. *Habitat International*, 22(3), 299-312.

Kowaltowski, D.C.C.K., Ruschel, R.C., Pina, S.G.M., Granja, A.D. and Oliveira, P.H. (1998). Examples of User Representation in Brazilian Housing and the Need for Professional Support in the Self-Building Process. In J. Tecklenburg et al. (eds.), Shifting Balances, Changing Roles in Policy, Research and Design. IAPS 15, Eirass European Institute of Retailing and Services Studies, pp. 54-65.

Kowaltowski, D.C.C.K., Fávero, E., Ruschel, R.C., Pina, S.A.M.G., Labaki, L.C., Borges Filho, F. (2000). Automated Design assistance for Self-Help Housing in Campinas, Brazil. In H. Turgut and P. Kellet (eds.), Cultural and Spatial Diversity in the Urban Environment. Istanbul: IAPS CSBE Network Book Series: 3, pp. 67-72.

Kowaltowski, D.C.C.K., Pina, S.A.M.G., Labaki, L.C., Borges Filho, F. (2002). A Basic Teaching Instrument For Awareness Of Environmental Comfort. In R. García-Mira, J.M. Sabucedo and J. Romay (eds.) Culture, Quality of Life and Globalization. Book of Proceedings, IAPS 17th. A Coruña: Asociación Galega de Estudios e Investigación Psicosocial.

Kowaltowski, D.C.C.K., Pina, S.A.M.G. and Ruschel, R.C. (1995a) Relatório Científico: *Elementos Sociais e Culturais da Casa Popular, Campinas-SP*. Faculdade de Engenharia Civil, UNICAMP, Campinas: FEC-UNICAMP.

Kowaltowski, D.C.C.K., Pina, S.A.M.G., Ruschel, R.C. and Oliveira, P.V.H. (1995b) *Relatório Científico: Uma Metodologia de Projeto Para a Casa Popular na Cidade de CAMPINAS-SP*. Faculdade de Engenharia Civil, UNICAMP, Campinas: FEC-UNICAMP.

Labaki, L.C. and Kowaltowski, D.C.C.K. (1998). Bioclimatic and Vernacular Design in Urban Settlements in Brazil. *Building and Environment*, 33(1), 63-77.

Nolasco, A.M. (1995). *Caracterização do processo de autoconstrução no município de Piracicaba/SP*. In Seminário Nacional sobre Desenvolvimento Tecnológico dos Pré-Moldados e Autoconstrução, Proceedings NUTAU, São Paulo: FAU USP.

Ornstein, S.W., Roméro, M.A. and Cruz, A. de O. (1995). *Avaliação funcional e do conforto ambiental de habitações autoconstruídas: o caso do Município de São Paulo*. Proceedings of the Seminário Nacional sobre Desenvolvimento Tecnológico dos Pré-Moldados e Autoconstrução, (pp. 249-259). USP, São Paulo: FAU USP.

Patarra, N., Baeninger, R., Bogus, L. and Annussi, P. (eds) (1994) Migrações, Condições de Vida e Dinâmica Urbana. Campinas: Instituto de Economía, UNICAMP.

Pettang, C. and Tatietse, T.T. (1998). A New Proposition for the curbing of spontaneous housing in urban areas in Cameroon. *Building and Environment*, 33(4), 245-251.

Rapoport, A. (1988). Spontaneous Settlements as Vernacular Design, in Spontaneous Shelter: International Perspectives and Prospects. Philadelphia: Temple University Press.

Schulz, C.(1996). *Pequeno Comprador Domina Mercado de Cimento* in O Estado de São Paulo, (newspaper). Caderno de Economia, 05.05.1996, São Paulo, 1996, pp. B9.

Tipple, G. (2000). Extending Themselves: User-initiated Transformations of Government Built Housing in Developing Countries. Liverpool: University of Liverpool Press.

Turner, J.F.C. (1976). Housing by People: Towards Autonomy in Building Environments. London: Marion Boyars Publishers Ltd.

Valença, M. (1999). The closure of the Brazilian Housing Bank and Beyond. *Urban Studies* 36(10), 1747-1768.

Valença, M. (1992). The inevitable crisis of the Brazilian Housing Finance System. *Urban Studies*, 29(1), 39-56.

Watrin De Rosa, V. (2003). *O Significado da Tradição na Autoconstrução de Moradias*, Masters Thesis, Faculdade de Engenharia Civil, Universidade Estadual de Campinas, Campinas: FEC- UNICAMP.

Werna, E. (1996). Business as Usual: small-scale builders and the production of low-cost housing in developing countries. Aldershot: Avebury Publishers.

Chapter 6

Assessing the Acceptability of Alternative Cladding Materials in Housing: Theoretical and Methodological Challenges

Anthony Craig, Leanne Abbott, Richard Laing and Martin Edge

Introduction

The external appearance of housing has an impact on both the quality of the immediate neighborhood in which it is constructed and the surrounding landscape. In the UK the priorities of government and speculative builders have often been at odds with the priorities and preferences of the final occupants of new housing. There are currently many calls for change in the house building industry coming from the UK government (Egan, 1998) and industry (Sparksman et al., 1999). The need for these changes has come about for a number of reasons. These include the skill shortage in the building trade (Clarke and Hermann, 2001) and an increasing emphasis on 'customer focus' (Barlow and Ozaki, 2000). However, there is a general conservatism among the various stakeholders in the house-building process. This conservatism often manifests itself in assertions that the house-buying public are resistant to change in the housing product (Lutzenhiser and Janda, 1999; Ball, 1996), and that therefore some form of market transformation would be required to make such changes acceptable to the public.

The particular change in house design that is examined in this paper is driven by a desire by many developers and others to increase the use of lightweight rainscreen cladding on the exterior facades of houses. At present, most timber frame buildings in the UK are constructed with an external masonry skin, which often serves only as a protective rainscreen cladding. There are various drivers influencing the recent emphasis on timber cladding, foremost of which is that of sustainability, along with various other cost and performance benefits that this technology might offer (Davies et al., 2002). As one timber cladding specialist put it: '*housebuyers are often resistant to timber cladding since they perceive timber-clad and timber-framed homes to be somehow inferior to "traditional" masonry-clad, timber framed houses*' (Davies et al., 2002). The idea that people perceive timber as being somehow inferior is often stated by developers, builders and a range of built environment professionals, but evidence for this tends to be anecdotal. The lack of empirical studies that evidence this claim was the starting point of this research.

As part of a larger study of potential resistance to change in the way we build our houses, this research looked at the effect of various external cladding materials on judgments made about detached houses. Various studies have previously shown there to be a relationship between the observable attributes of building exteriors and building preference (Herzog and Shier, 2000) and perceptions of housing quality (Reis, 2001). Further studies have explored the socio-psychological processes which contribute to housing preferences, which may manifest in ratings of the emotional qualities of the building itself (Nasar, 1994) and the social identity of potential occupants (Sadalla et al., 1987). Sadalla and Sheets (1993) argue that the materials from which houses are constructed convey more meaning to people than simply the physical properties of the materials. They argue, through a series of studies, that building materials employed on exterior facades have a function in defining the social identity of home-owners (Sadalla and Sheets, 1993), as well as having functional utility in themselves. So, for example, someone actively choosing to live in a timber-clad house might be rated (by others as well as themselves) as more artistic, or less conservative than those choosing an exterior facade of concrete block, if these were the personality characteristics attributed to these particular materials by the rater. It should be noted however, that Sadalla et al. (1987) found that interior cues were seen as a more informative measure of social identity than exterior cues, possibly due to the increased flexibility people have concerning the manipulation of interior cues.

Given the limited choice however, preferences for external cladding materials will tend towards whatever is dominant within the housing environment. For people to be able to employ the symbolic aspects of materials in the process of defining social identity (Sadalla and Sheets, 1993) would require an environment within which choice of materials for exterior facades is exercised in reality. While there are clear geographical variations in Britain in the use of different exterior cladding materials, there has nevertheless been a historical tradition and preference towards brick in England, and roughcast in Scotland. Although innovation in the external appearance of dwellings was encouraged during the 1950s and 1960s (Brindley, 1999), the cleavage between the architectural and social discourses after that time arguably led to the ghettoization of many housing estates of 'modern' design, thereby stigmatizing the use of innovative approaches to housing design. It would be interesting to know the extent to which any negative evaluations of a particular material are tenure-specific, and if so what the social effect would have been of such mass-innovation within the private housing sector. Although there is some evidence to suggest that timber cladding has a long history in Scotland (Edge and Pearson, 2001), it is nevertheless arguably not embodied within the collective memory, and is still considered by many to be in the class of 'unfamiliar'. Social representations (Moscovici, 2001) are the concepts shared as 'common sense' by members of a collectivity. According to Moscovici, the role of these social representations is to conventionalize something, or locate it in a familiar context. The social representation of timber cladding in housing (if indeed one exists) is likely to be responsible for resistance or rejection of this material, if the established order, or what is *familiar*, is perceived to be in threat (Moscovici, 2001). This would suggest a need for a new representation of this particular innovation within society if it is to become acceptable and 'familiar'.

Linking this with the idea that meanings associated with building materials are employed in the defining of social identities (Sadalla and Sheets, 1993), it is likely that due to this stigmatization of 'non-traditional' cladding materials, they will be associated with 'low-status', and are hypothesized therefore likely to be rated as less pleasant and worthy of purchase consideration than claddings such as brick and roughcast.

A study by Taylor and Konrad (1980) found that people (a Canadian sample) were generally unsupportive of the idea of a '*disposable urban fabric readily replaced in the cause of change*' (p305), but rather, tended to be strongly inclined towards the preservation of the past. The political impetus for change in the UK away from what is perceived as part of 'the past' towards something perceived as 'new' may receive a similar lack of support. It is interesting to note that the housing stock in Britain is fairly old in comparison to many countries, with 48% having been built before 1945. In the Netherlands, 47% of the housing stock has been replaced since 1971, compared with only 21.8% in Britain (Clarke and Hermann, 2001). It is quite likely that this fact will have a significant impact on the judgments people make about so called 'innovative' cladding materials, as these are likely to be made by way of comparison with what is generally thought of as an 'average house'. Therefore, if most houses a person is familiar with are clad in brick, and these houses also appear to be of a significant age, then it is hypothesized that the material brick is likely to be attributed with such qualities as 'tradition' and 'durability' as well as familiarity.

Method

Participants

844 people took part in this study, 49.5% of whom were male. Each respondent made judgments on six houses (out of a total of 36), therefore 5064 sets of judgments were made in total (see below). Responses within the local area were via a hand-delivered postal survey, whereas responses for the rest of the UK were via an internet survey (16% of total).

Stimulus Materials

Six cladding materials (brick, roughcast, precast concrete, vertical timber, horizontal timber, and painted horizontal timber) were included in the study materials, following a successful piloting exercise. Two roofing materials (slate and corrugated steel sheet) were also varied, along with three house-design (1, 1½ and 2 storey houses selected from a range of timber-framed houses currently on the market). Therefore, there were 36 total variations of cladding, roofing, and house type (6 x 2 x 3 = 36).

Images were generated by modelling each of the houses in AutoCAD®, and then rendering each in 3D Studio MAX®. These images were then pasted as a layer into two Photoshop® images of a photograph (of a site for a detached house in Aberdeen, Scotland). The background for all of the houses was the same, so as to reduce bias introduced by the immediate setting. As there are two views of each house, this meant that the total number of images generated was **72**. These images were then piloted to check that the images were judged to be realistic representations of the cladding materials being modelled. The results of this pilot study were satisfactory with a high degree of recognition of wall and roof cladding type and hence the images were included in the surveys. All images presented in the surveys were of near-photographic quality. Indeed, in the pilot study, most people thought the houses depicted were actually real houses. An example of the images shown in the survey, but in monochrome as opposed to the original colour, can be seen in Figure 6.1.

Figure 6.1 Example images from the survey (actual size used for the paper-based survey was 113 x 85mm, and all images were in colour)

There were seven different survey instruments in total, as each house needed to be judged on 10 items, and it was felt that the survey would be too onerous for each respondent to cover all 36 variations. The presentation order of the house variations was randomised for each survey, so as to control for order effects as much as possible. In order to amalgamate the results of the seven surveys, responses to each house presented are treated as if they were from a separate person, therefore the number of respondents in the analyses is effectively multiplied by 6 (844 x 6 = 5064). Hence, all subsequent analyses will assume a between-subjects design.

Measures

Each of the houses were rated on the following 10 seven point Likert-type scales, from strongly-agree to strongly-disagree:

1. I would consider buying this house
2. This house has a pleasant appearance
3. The house style is 'traditional'
4. This house strikes me as being unusual
5. This house strikes me as being boring
6. The colours of the materials complement each other
7. The house looks like it will last a long time
8. I find this house unappealing
9. I would say the house style is modern
10. I think developers could easily sell houses like this

Demographic information was also recorded, as well as any further open-ended comments respondents wanted to make about each house.

Procedure

Respondents each received one of the seven surveys. 2728 surveys were hand delivered to a cross section of dwellings in the Aberdeenshire area, and returned in a freepost envelope provided. An overall response rate of 26% was achieved for the paper survey, with a total of 708 surveys being returned by post. It is almost impossible to determine a response rate for the internet based survey, as it is difficult to know how many people did not do the survey but were aware of its existence. A total of 136 surveys were completed over the internet, after screening out surveys which were completed incorrectly. The survey took approximately 15 minutes to complete, and a great deal of interest was generated, as reflected in respondents' comments.

Results

An initial check was performed on the data to test the hypothesis that brick was likely to be attributed with such qualities as 'traditional' (q3) and 'durability' (q7). As shown in Figure 6.2, brick is rated as the most traditional (closely followed by roughcast), and also the most 'long lasting' of all the cladding types.

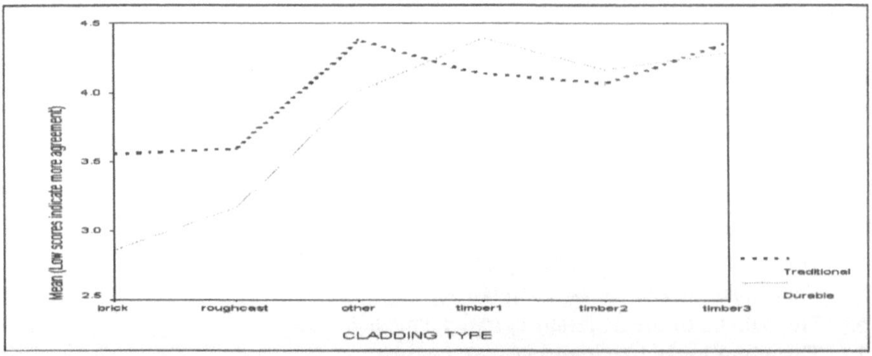

Figure 6.2 **Effect of cladding material on perceptions of tradition (Brick/Timber2 – Mann-Whitney U: z = 6.38; p<0.05) and durability (Brick/Roughcast – Mann-Whitney U: z = 4.00; p<0.05)**

A check was also carried out to see if 'non-traditional' cladding materials were rated as less pleasant and worthy of purchase consideration. As Figure 6.3 shows, in general this was confirmed for both, although timber 2 (horizontal timber cladding) stood out in both cases as both more pleasant and also more worthy of purchase consideration than the other 'non-traditional' claddings. In fact, the ratings of pleasantness for brick and timber 2 were not statistically significant.

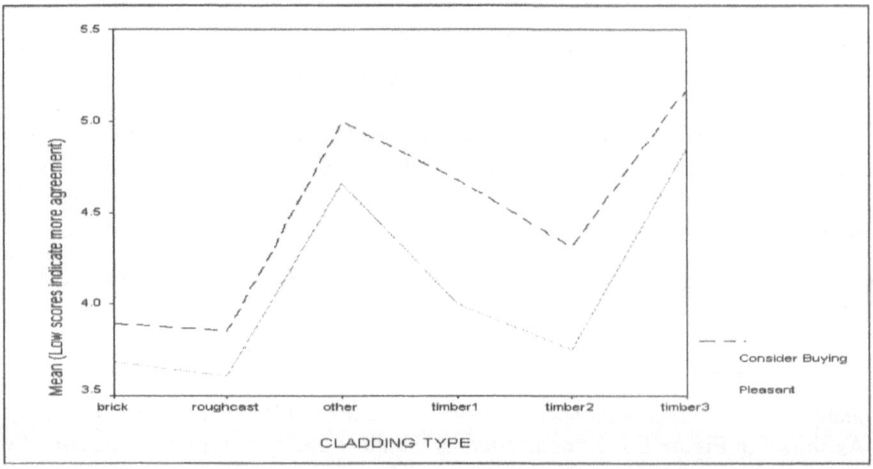

Figure 6.3 **Effect of cladding material on perceptions of pleasantness (Brick/Timber1 – Mann-Whitney U: z = 3.46; p<0.05) and purchase consideration (Brick/Timber2 – Mann-Whitney U: z = 4.37; p<0.05)**

In order to come to an overall assessment of each house-variation, 6 of the items were combined into a scale (items 1,2,3,6,7 and 10). A reliability analysis on these items produced a Crombach's alpha of 0.88, and as such the scale was taken to be an acceptable measure of preference for the purposes of this study.

A one-way ANOVA with cladding as the between subjects variable revealed that the type of cladding was a significant predictor of preference ($F_{5,4845} = 109.2$, $p < 0.01$), as measured by the scale described above. This effect was examined in more detail by contrasting each of the claddings with one another using t-tests. With the exception of the contrast between 'brick' and 'roughcast' ($t = -0.171$, $p > 0.05$), all other contrasts were significant (see Table 6.1).

Table 6.1 Effect of cladding material on preference

Cladding	Means	S.D.	$F_{(5,4845)}$	Contrast t-tests
Brick	20.84	7.95	109.2**	Brick and Roughcast (t = -0.171)
Roughcast	20.91	8.63		Roughcast and H-Timber (t = -6.99)**
Precast Concrete	26.42	7.94		H-Timber and V-Timber (t = -3.74)**
Vertical Timber	25.25	8.15		Precast and V-Timber (t = 2.94)**
Horizontal Timber	23.77	7.80		Precast and P-H-Timber (t = -4.47)**
Painted Horizontal Timber	28.20	8.04		

Preference Score: potential range from 6 (most positive preference) to 42 (most negative preference). ** p<0.01

Thus, the order of preference for cladding materials (starting with most preferred) is: brick, roughcast, horizontal timber, vertical timber, precast concrete, and then painted horizontal timber.

The influence of roofing materials and house type was also examined through similar analyses (see Table 6.2). Overall, slate was preferred to steel ($F_{1,4849} = 316.43$, $p < 0.01$), and the order of preference for house type (starting from most preferred) was: 1½ storey, 2 storey, and then 1 storey ($F_{2,4848} = 56.318$, $p < 0.01$).

Table 6.2 Effect of roofing and house type on preference

Roofing	Means	S.D.	$F_{(1,4924)}$	
Slate	22.14	8.59	316.43**	
Steel	26.37	7.91		
House			$F_{(2,4848)}$	Contrast t-tests
1 Storey	25.95	8.21	56.316**	1 Storey and 2 Storey (t = 6.82)**
1½ Storey	22.84	8.65		2 Storey and 1½ Storey (t = -3.70)**
2 Storey	23.96	8.43		

Preference Score: potential range from 6 (most positive preference) to 42 (most negative preference). ** p<0.01

A further ANOVA was carried out in which all three factors (cladding, roofing, and house type) were entered into a factorial analysis. A quarter of the preference variation could be accounted for by the independent variables ($R^2 = 0.245$). All main effects and interactions came out as significant at the $p<0.05$ level of significance, including a significant main effect of cladding material on preference ($F_{5,4815} = 153.83$, $p < 0.001$), which means that even when roofing and house type are taken into account, cladding material is still a significant predictor of preference score as measured in this survey.

Methodological Considerations

In addition to its primary purpose, this study aimed to test some methodological innovations, employing the use of state-of-the-art computer modeling and internet-based survey methods, which used high quality visual images to explore perceptions of the built environment.

The use of sophisticated computer modeling techniques provided the means by which the surrounding environment in which the houses are viewed could be controlled for, and therefore eliminated in the analyses as a source of response variance. Of course, there may be some residual response variance due to perceptions of how well houses are judged to fit in with the surrounding environment. However, as the houses presented in this study are all fairly similar, it was assumed that this effect would be negligible. This study found that such computer modeling techniques can produce a high degree of realism, as demonstrated by respondent comments to the effect that they thought the houses were in fact real houses. Similar techniques have been used in a variety of other experiments in recent years, mainly concerning the visual impacts of streetscapes (e.g. Davies and Laing, 2001; Al-Kodmany, 1999), and have been found to be extremely useful in gauging preferences in the built environment.

Whilst a large amount of respondents in this study participated in the survey through a fairly conventional paper-based questionnaire, a substantial number participated via the internet. Internet surveys are becoming an increasingly popular method of data collection in the social sciences. Indeed, if conducted carefully, the advantages of web-based research can by far outweigh the disadvantages, although there is an '*alarming potential for configuration errors*' (Reips, 2001), which can lead to bias or misleading results. One of the many advantages of such methods is the ability to gain access to geographically distant populations at much lower costs than traditional paper-based surveys. Also, the data from such studies can be stored in an online database (in this case MySQL), which eliminates the need for data transcription. Among the various disadvantages are self-selection bias, potential for multiple submissions, and lack of experimental control (Reips, 2001). The issue of experimental control is particularly important for surveys such as the one in this study, involving the presentation of images, as it is very important to know that all respondents see the same stimulus materials. Screen size and resolution is arguably the main issue in this respect, which was controlled for as far as possible by designing the survey to work on a small screen size. The issue of download time also has to be traded off with image size in such surveys, as it is known that respondent drop-out is increased dramatically if download speeds are low (Dillman, 1999).

Another advantage of using web-based technologies for carrying out surveys like this is the ability to randomize both the presentation order, and the survey itself. This was particularly useful in this survey, as it was important that all 36 house type variations were equally covered for responses. Potentially, if any particular survey needed more responses relative to the others, it would then be possible to target respondents with that survey if need be (although this wasn't necessary in this study). Results so far suggest that there are no major differences in the results when comparing the internet sample and the paper-based sample that are not potentially explained by geographical variations or other demographic factors.

Discussion

The results presented here support the hypothesis that exterior cladding materials have a significant impact on overall housing preference. More specifically, respondents in this survey rated brick and roughcast as being more durable and traditional than all the other cladding materials presented. Respondents also rated these two cladding materials as being more pleasant, and more worthy of purchase consideration than the other materials, with the exception of horizontal timber cladding. In terms of roofing material, slate was preferred to steel. These results suggest a general disposition within the sample studied towards what are generally thought of as being 'traditional' exterior cladding materials, confirming the hypothesis presented in the introduction.

Overall, it would appear then, that some of the anecdotal evidence given by developers and builders to the effect that house-buyers prefer 'traditional' cladding materials is now supported by empirical evidence. Furthermore, it seems that if there is any evidence for a latent collective memory for timber cladding coming through from the results, it is specific to horizontal timber cladding, which was rated as more traditional than the other two timber claddings, and also more worthy of purchase consideration. This is interesting insofar as house-builders specifying timber claddings for houses today would probably be more likely to opt for the vertical timber for rational technical reasons. So arguably, specifying the technical optima in terms of timber cladding would be predicted to have an adverse impact on acceptability if the results of this study were generalized to the wider population. The reasons for this finding will be studied in the later stages of this project.

The finding that respondents rated as more pleasant and worthy of purchase those materials that were also rated as more traditional provides support for the application of social representations theory to market research in housing, and possibly in a wider context to purchase considerations in general. This is interesting in that it suggests a possibility that social-psychological theory and research can be useful in highlighting the reasons for a variety of consumer behaviors. Understanding these behaviors is important not only for the more obvious reasons of knowing what sells, and therefore what to sell. Consumer behavior is also a key indicator as to what values people hold, and this is particularly important within the context of sustainability. With this in mind, it would be useful to examine the impact of making salient the sustainability aspect of timber cladding, and whether or not this would be reflected positively in people's evaluations of such houses.

Acknowledgements

This research was carried out as part of a larger project funded under the MCNS 'LINK' Program, by the Engineering and Physical Sciences Research Council (EPSRC), and the Department of Trade and Industry (DTI). We would also like to thank Stephen Scott, who was responsible for producing all of the computer images used in this study.

References

Al-Kodmany, K. (1999). Using visualization techniques for enhancing public participation in planning and design: process, implementation, and evaluation. *Landscape and Urban Planning*, 45, 37-45.

Ball, M. (1996). Housing and Construction - A Troubled Relationship?, Joseph Rowntree Foundation. Bristol: The Policy Press.

Barlow, J. and Ozaki, R. (2000). User Needs, Customisation and New Technology in UK Housebuilding, *ENHR 2000 Conference, Gävle, Sweden*.

Brindley, T. (1999). The Modern House in England. In T. Chapman and J. Hockey (eds.), *Ideal Homes? - Social Change and Domestic Life*, (pp. 30-43). London: Routledge.

Clarke, L. and Hermann, G. (2001). Innovation and Skills: A Transnational Study of Skills, Education and Training for Prefabrication in Housing. IMI Project, Final Report Westminster Business School.

Davies, A. and Laing, R. (2001). *Streetscapes: Their contribution to wealth creation and quality of life*, Final Report to Scottish Enterprise, August.

Davies, I., Walker, B. and Pendlebury, J. (2002). *Timber Cladding in Scotland.* Edinburgh: ARCA Publications Ltd.

Dillman, D. (1999). *Mail and Internet Surveys: The Tailored Design Method.* New York: John Wiley and Sons, Inc.

Edge, M. and Pearson, R. (2001). Vernacular Architectural Form and the Planning Paradox: A Study of Actual and Perceived Rural Building Tradition. *Journal of Architectural and Planning Research,* 18 (2), 91-109.

Egan, J. (1998). *Rethinking Construction*, Report of the Construction Task Force to the Deputy Prime Minister, HMSO July 16th 1998.

Herzog, T. and Shier, R. (2000). Complexity, Age, and Building Preference. *Environment and Behaviour,* 32 (4), 557-575.

Lutzenhiser, L. and Janda, K. (1999). *Residential New Construction: Market Transformation Research Needs.* CIEE Market Transformation Research Scoping Study (http://ciee.ucop.edu/docs/res_new_cons.pdf).

Moscovici, S. (2001). *Social Representations: Explorations in Social Psychology.* New York: New York University Press.

Nasar, J. (1994). Urban Design Aesthetics - The evaluative Qualities of Building Exteriors. *Environment and Behaviour,* 26 (3), 377-401.

Reips, U. (2001). *Internet-Based Psychological Experimenting*, Keynote Presentation at the first Psychology and the Internet conference, The British Psychological Society, Farnborough, 7th-9th Nov 2001.

Reis, A. (2001). *Housing appearance as an indicator of housing quality*, Proceedings of the 32nd Conference of the Environmental Design Research Organization (EDRA), Edinburgh, 68-74.

Sadalla, E., Vershure, B. and Burroughs, J. (1987). Identity Symbolism in Housing. *Environment and Behaviour,* 19 (5), 569-587.

Sadalla, E. and Sheets, V. (1993). Symbolism in Building Materials - Self Presentational and Cognitive Components. *Environment and Behaviour,* 25 (2), 155-180.

Sparksman, G., Groak, S., Gibb, A. and Neale, R. (1999). *Standardisation and pre-assembly: adding value to construction projects*, CIRIA Publications: Report 176.

Taylor, S. and Konrad, V. (1980). Scaling Dispositions Toward the Past. *Environment and Behaviour,* 12 (3), 283-307.

Chapter 7

Neighbourhood Quality of Life – Global and Local Trends, Attitudes and Skills for Development

Ombretta Romice

Research on Neighbourhood Development

For their size and impact on daily life, neighbourhoods are ideal units to study and assess quality of life because they combine physical and social scale: social networks tend to overlap to spatial arrangements and issues of territoriality, identity and well-being become attached to location (Moudon, 1994; Morrison, 2003). Whilst it is acknowledged that neighbourhoods act as important sources of opportunity and provide a sense of identity, on the other hand, they can also act as a constraint on personal life chances (Madanipour et al. 1998).

Over its three years of work, an international consortium under the acronym of NEHOM, of which the author was postdoctoral researcher, has studied deprived neighbourhood throughout eight European counties (UK, France, Italy, Sweden, Germany, Hungary, Estonia, Norway) to understand the effectiveness of housing policies, physical, social and economic initiatives in relation to phenomena of social exclusion, segregation, stigmatisation and quality of life [EU Funded Project 'Evaluating housing and neighbourhood initiatives to improve the quality of life of deprived urban neighbourhoods and assessing their transferability across Europe' [Acronym: Neighbourhood Housing Models (NEHOM)], Work programme: EESD, 1.1.4.-4.1.2 Improving The Quality Of Urban Life].

The focus of the research are socially excluded neighbourhoods and how they afford - or not - personal life chances, and how can their affordances be enhanced through initiatives and programmes investing in the reconstruction of social dynamics as a catalyst of further improvement. The underlying belief and driver of the research was that residential based networks are the building blocks to social cohesion (Castells, 1997) and that the nature and quality of social interaction at the local level is a condition for social cohesion at societal level.

In total the team examined 29 neighbourhood types, from large panel-housing estates, generally post WW II (13 studied), inner city districts of turn of the century fabric, (12 studied) to fairly central neighbourhoods (4 studied)[1]. During the first 18 months research was carried out through objective investigation (Census, social, economic, physical surveys) and structured in-depth interviews with residents, local leaders and those involved in the management of the neighbourhood housing stock and in the formulation of housing and development policies for the neighbourhood[2]. The aim of this investigation was to establish the quality of living conditions (past and present) as perceived by those interviewed, as well as through publicity and literature about the neighbourhood. The following 18 months were devoted to the search for those factors and initiatives - physical, social, political, economic and cultural - that allowed an improvement of the quality of life in these neighbourhoods; overall aim of this second phase was to complete a collection of organisational principles and operative details for the instruction of EU, national and local housing authorities, grassroots organisations etc.

The level of integration of these neighbourhoods to the economic, political and structural city-core varies, as does their size, physical condition, social composition, degree of organisation and management. They are, case-by-case, agglomerates of distinctive character marked by geographical boundaries; spatial and organisational units relying on self-contained local resources; or revolving around an initiative of significance to its residents and local organisations. It is therefore difficult to draw parallels between these initiatives. Still, the 29 neighbourhoods share similar aspects. If one accepts that residential preference is a factor determining residential location patterns (Adams and Gilder, 1976) as well as economic and cognitive resources (Van Kempen and Ozueken, 1998; Murie, Knowrr-Siedow and van Kempen, 2003), then in most of the cases these are the neighbourhoods where people live when they don't have the possibility to move elsewhere; or those where people have been living for long enough to feel an identity with the area and want to stay despite its poor quality; or where residents consider this accommodation as transient.

The number of such neighbourhoods is significant throughout Europe for historical reasons (demographic pressure, war destructions, emigration in pursue of employment) and in many cases they host the poorest pockets of society and recently, in certain cases, large numbers of immigrants searching for employment in western economies. Despite the changed situations these neighbourhoods are extremely significant and might hold a key to understanding how to achieve and maintain social cohesion and governance, intended as a complex balance of interests and actors, processes and mechanisms (Vicari Haddock, 2004), in parallel

---

1 All case studies can be viewed in Holt Jensen et al. (2004 eds). *New Ideas for Neighbourhoods in Europe*. Tallinn: TUT Press.

2 Each case study included around 20 in-depth interviews with local residents and around six interviews with actors involved in policy and service delivery. Although a representative sample was not intended, it was important that the local residents interviewed included people that were at risk of social exclusion, such as long term unemployed, lone parents, elderly and ethnic minorities.

and through their process of urban renewal. The European Community and individual states have acknowledged that neighbourhoods are the key spatial scale for policy intervention and the point around which coordinated action for urban regeneration could revolve. Contemporary cities are a mosaic of functionally defined neighbourhoods; understanding how they work, and the neighbourhood structure of cities (land uses, the socio-economic groups that populate the neighbourhood, its connections to other neighbourhoods, and externalities that affect it and that spread from it) is paramount to any action of regeneration (Kirwan, 1996:50).

Local Dynamics and Problems to Address

Although it is difficult to generalise, it is noteworthy that many that these neighbourhoods have witnessed an escalation of problems during the 1980s and 1990s, when poverty was growing throughout Europe for a number of related causes. Three areas where substantial change has taken place since the 1980s can help explain how entrenched these problems are, and the difficulty that any system or initiative is bound to encounter in addressing them: modification in the labour market, in the family structure and in the provision of State welfare policies.

In addition, almost all the neighbourhoods studied are facing emerging trends - a direct manifestation of processes of globalisation, and in particular of further changes in the labour market conditions (Holt-Jensen, 2002). Growing waves of labour emigration and political refugees are creating new social patterns, polarising even further society through relative wealth and ultimately changing the socio-physical geography into pockets of spatial segregation, where poverty coexists with racial tension.

The co-presence of several or all of these trends risks generating long-term incapacity to satisfy basic needs, with an effect on individual capacity to establish and maintain social networks, causing 'disaffiliation' and difficulty in accessing social services (Castel, 2000). Similarly, Vicari Haddock (2004:126) calls this 'careers in poverty', the situations in which individuals cannot face negative events and enter worsening conditions. Holt-Jensen suggests Castel's idea that these socio-economic processes have two outcomes: a structural one, the rich get richer, the poor get poorer and the middle-class is shrinking; and a spatial one, the segmentation of economic exclusion, more dramatic in those neighbourhoods built to accommodate the workers of now closed manufacturing industries. The spatial distribution of these areas of poverty changes widely, but our studies indicate that before the initiatives started they shared some commonality. Each was located in estates spatially isolated from the city core and in central areas but under pressure from phenomena of gentrification. In addition mobility opportunity, including access to jobs and services was much reduced.

Still, the NEHOM study of the relationships between states, housing policies, and initiatives against social exclusion, suggests clearly that the description of these negative trends are far from a description of reality, and that much is happening that shows, at different scales, how such trends can be reverted. Urban life is a delicate balance between employment, family, leisure and spiritual requirements. New factors (other than the traditional, such as job opportunities, economic capacity, mobility) that play a part in balancing these requirements are emerging and are strongly linked to organisational, social and communicational capacity within the neighbourhoods and between the neighbourhood and political/administrative agencies. Understanding their value is the key to understanding the functioning of these neighbourhoods and their opportunity of development as responsive and supportive environments for their residents.

Overall, the specific problems affecting the perception of life quality in the 29 neighbourhoods studied were:

- Social: high concentration of fragile groups; low level of residents' resources/ unemployment; lack of social interaction; feelings of insecurity; negative neighbourhood image;
- Physical: poor physical quality of the housing stock; lack of maintenance of outdoor spaces; vacant housing and building plots; physical isolation of the neighbourhood from the rest of the city;
- Managerial: inefficient housing management; low level of residents' participation; lack of private investment).

What Affects Quality of Life in Deprived and Socially Excluded Neighbourhoods?

The perceptions of quality of life in the 29 neighbourhoods were studied through in-depth investigation of the neighbourhoods, their population, those working in them and involved in their management, servicing and planning. Further investigation of the initiatives set in place to address these problems suggested that the initiatives best positioned to tackle them were those counteracting the fragmented institutional sector in its capacity to provide for those in need, and at the same time enabling a bottom-up development capable of reducing dependency, developing self-sufficiency and organisational capacity to disentangle these neighbourhoods from a one-way dependency on institutions and reinforce their delivery capacity. Some of the general aims of these initiatives are described below.

- Social: Create new opportunities (social, cultural and economic) for existing residents groups, attract better off groups to the neighbourhood; facilitate dispersal and integration of fragile groups; improve level of residents' financial resources and skills; enable employment and local business development; encourage and enable community capacity building; reduce and control conflicts between residents; reduce and counteract feelings of insecurity in the neighbourhood; revert the negative image of the neighbourhood; attract private investors.
- Physical: Housing rehabilitation; upgrade of open spaces; maintenance and upgrade by residents; functional changes – land + buildings; turn vacancies

into housing for special groups; change ownership structure; counteract physical isolation of neighbourhood from surroundings.

- Management: Reorganise management system, develop coordinated partnerships/programmes; focus agenda on residents' needs and local conditions; encourage residents to take part to housing and neighbourhood management, formalise their role; create conditions to attract private investment.

All the initiatives investigated are concluded or have been in place for a number of years (the visible impact of the quality of life has a long life span). Similarly, their effectiveness was researched on several levels and did not necessarily focus directly on the objectives that it set to achieve.

Although it was clear that each of these actions was in a position to counteract a variety of problems and changes, one still has to question how sustainable these efforts are in the long run; how much individual initiatives can do in this broad network of urban transformations; and how relevant or significant these initiatives and the lessons we learned from them can be for other circumstances? According to the most recent studies on sustainability, the link between local and global sustainability is currently not supported by adequate knowledge and practice (Frey, 2004). Without such coordination, the efforts to generate a sustainable urban form, from a social, economic, physical and environmental point of view are in vain and not viable in the very near future (*ibid.*).

Perhaps the most interesting long-term objective of the successful programmes studied is to change the traditional relationship between Government and society as a whole, establishing new functions that can be carried out by citizens through social commitment and new forms of interaction between the two. Encouraging governance means accepting localised processes, where authorities (local, federal governments) become facilitators or coordinators, reducing dependency and encouraging empowerment and local connections within the neighbourhoods/districts and outside.

By concentrating on reducing residential turnover, mixing and stabilizing population, investing in the local economy, promoting social responsibility, creating safe environments, these programmes are working to create social cohesion (a form of social solidarity, which contributes to socio-economic security, an important alternative to respond to the changes that have occurred in family composition and in the state provision and support) and social capital. This, in turn, should lead to social citizenship, supplying the framework to fulfil one's right to enjoy a minimum level of material welfare.

Within the context of social housing and social exclusion (and related policies and initiatives), it is fundamental to understand that no single action can prove effective and lasting in enhancing quality of life in these neighbourhoods. Success is achievable through the integration of programmes, tools and efforts. Quality must be achieved through the reconstruction of confidence and capacity of their populations, and through the connection with mainstream initiatives and policies.

In particular, the experiences studied seem to prove that such integrated policies:

- Reduce dependency from unique, often one-off resources, because they draw together different sectors and areas such as social, economic, physical for which often funds are allocated. The resource-pool increases and offers more realistic opportunities for investment.
- Are self-generating - investment in one sector can generate spin offs in other sectors (in several of the projects studied for example, renewal work (housing stock, open space, community facilities) in the neighbourhood was contracted to companies which in turn trained the local unemployed. These then had the skills to set up local management groups for the progressive maintenance and repair of the stock in the neighbourhood).
- Improve the number and quality of connections and organization within a neighbourhood and between the neighbourhood, its outskirts, and local and central government, injecting an European dimension to it which can be not only a source of funding, but of experience, knowledge transfer and quality assurance.
- Considering the neighbourhood as part of a greater social network:
  - Increases social citizenship and cohesion, and ultimately political effectiveness. In turn this enhances the leasing power of the neighbourhoods in its demands for a quality environment.
  - Improves exchange, social, cultural and economic, reducing actual abd a culture of dependency from state provision/assistance.
  - Reduces isolation, and in the long term can reverse a negative neighbourhood image.
  - Increases the degree of dependability on the neighbourhood, improving its significance in overall processes of urban regeneration.
  - Adds an outward dimension to a local focus.
  - Enhances the potential of local entrepreneurship.

All 29 NEHOM initiatives studied have pursued similar ambitious goals (specific to national conditions and trends). This does not mean that their effectiveness can be guaranteed in relationship to long-term development: experience, well established links and procedures, good balance between centralization and decentralisation, procedural clarity and flexibility are practices hard to learn and establish. In all cases though, the residents' perceived quality of life in the neighbourhood has improved during or after the completion of the initiative itself. The overall quality of life in the neighbourhood was related to an improvement/increase in the economic support offered to families, in the range of services and facilities available, in the social and personal support offered to individuals, families and/or groups; in the offer of training and job opportunities; in the physical quality of the private dwellings and public open spaces; in the management process of the housing and other built stock, open spaces and play facilities.

Amongst the clearest outcomes of this research is the need for a well developed profession capable of managing and integrating the issues listed above together. First of all, it should possess and maintain an overview on international globalisation trends, national policies and mechanisms, local agencies and activism, and establish connections and dependencies between levels. At the same time it should attempt to ensure that such connections and dependencies are not unique. Then, it should be able to generate (local) self-growth and development, strengthening local capacity to interchange links and connections according to needs, pressures, and changes in local dynamics.

Finally, it must acknowledge the value of integrated neighbourhood development and pursue it combining experience with awareness in other successful fields. Most of all though, whoever will be called to manage and deliver integrated development must acknowledge that, despite having been on the agenda and debated for over 10 years now, integrated development and its practice can never rest, being by nature dependent on change and circumstances. The acknowledgement of these potentials should be the first challenge of this new profession.

Are We Up for This?

The role of a university and academic department in relationship to urban problems is potentially as active as ever since the inception of the community design movement in the United States and in Europe in the 1960s and 70s. Design education, practice and research address increasingly complex questions, systems and problems through a synthesis across disciplines. Even more so does education for the design of the built environment in its aim to match human needs and aspirations to the scale and spatial quality of the built environment. Several are the disciplines that can enhance good place making – not only design-based disciplines, such as architecture, planning, landscape and urban design, but also disciplines centred around social studies, such as sociology, psychology, geography.

Romice and Uzzell (forthcoming) have for over three years now worked to reinforce the link between design and environmental psychology. No matter how valued the results of such attempts, the collaboration between the two is still very limited, the reciprocal strengths have not been fully exploited and there are still ideological and practical barriers that inhibit the professional relationship. Still, it is clear enough that both disciplines (i.e., architecture and psychology) have much more that an academic curriculum to offer each other: they have theoretical beliefs, practical skills, analytical frameworks and simply 'ways of seeing' whose combination could produce exponential benefits rather than simple additive increments to the competencies of both.

Successful living environments are clearly no longer solely a matter of good design: they must integrate cultural, entrepreneurial, social politic and civic sphere with the physical one. Graduates from both fields work in design practices, become consultants, politicians, community workers and much more. Education must acknowledge this development and be prepared for it. Personally, as an architect involved in design education who has had the experience of research on the quality of life across European neighbourhoods, I am working to redesign the curriculum of my classes and my studio work to expose students to this much broader set of issues, to alleviate the traditional disconnection between architecture, research and other disciplines. Architecture academic departments for example constitute only a small fraction of the total discourse on architecture and I suspect that academics exercise far less power in the field than academics who occupy a similar position vis à vis their practicing colleagues. in other disciplines. Architecture is little influenced by the academic world with the main journals of architecture often being disconnected from academic production; thus intellectual influences rarely penetrate architecture (Stevens, 1998).

Design is once again strongly embedded in the social sphere. However, the difference is that designers must now acknowledge that they are only one part of broader development and that to play their part successfully they must learn and collaborate more effectively and willingly. They must also become more comfortable with the idea that design is no longer a panacea for urban problems, but an intelligent and substantial complement to address it. To address issues of quality of life and in particular the life in deprived neighbourhoods, actions must be based upon an overview of international globalisation trends, national policies and mechanisms, local agencies and activism, and establish connections and dependencies between levels, while at the same time making sure that such connections and dependencies are not unique. Then they should be able to generate (local) self-growth and development, strengthening local capacity to interchange links and connections according to needs, pressures, changes in local dynamics.

Finally, architects and urban designers must acknowledge the value of integrated neighbourhood development and pursue it combining their experience and awareness of other cognate fields. Whoever is called upon to manage and deliver integrated development must acknowledge that, despite having been on the agenda and debated for over ten years now, integrated development and its practice can never rest, being by nature dependent on change and circumstances.

References

Adams, J.S. and Gilder, K.S. (1976). 'Household location and intra-urban migration'. In D.T. Herbert and R.J Johnston (eds.), *Social Areas in Cities, Spatial Processes and Form* (1) (pp.159-192). London: Wiley.

Castel, R. (2000). The roads to disaffiliation. Insecure work and vulnerable relationships. *International Journal of Urban and Regional Research*, 24(3), 520-535.

Castells, M. (1997). *The power of identity*. Oxford: Blackwell.

Frey, Hildebrand W. 2004, 'The Search for a Sustainable City. An Account of Current Debate and Research', paper from the 21st Conference on Passive and Low Energy Architecture, Eindhoven.

Holt-Jensen, A. (2002). The 'Dual City Theory' and Deprivation in European Cities. Paper presented at the 'XVI AESOP Congress in Volos', Greece July 2002.

Holt-Jensen et al. (2004 eds). *New Ideas for Neighbourhoods in Europe*. Tallinn: TUT Press.

Kirwan, R M. (1996). *Strategies for housing and social integration in cities*. Organisation for Economic Co-operation and Development. Paris: OECD Publications and Information Center.

Madanipour A, Cars, G. and Allen J. (1998 ed). *Social Exclusion in European Cities*, London: Jessica Kingsley Publishers.

Morrison, N. (2003). Neighbourhoods and social cohesion: Experiences from Europe. *International Planning Studies*, 2 (8), 115-138.

Moudon, A. V. and Ryan, M. (1994). Reading the residential landscape. In S.J. Neary, M.S. Symes and F.E. Brown (eds.), *The Urban Experience. A People-Environment Perspective* ,(pp. 183-313). Suffolk: St Esmondsbury Press.

Murie A, Knowrr-Siedow T. and van Kempen R. (2003). *Large Housing Estates in Europe. General Developments and Theoretical Backgrounds*. The Netherlands: A-D Druk bv, Zeist.

Romice, O. and Uzzell, D. (forthcoming) *Forty Years On: A Capital Experience in Psychology/Architecture Collaboration*.

Stevens, G. (1998). *The Favored Circle*. Cambridge: MIT Press.

Van Kempen R. and Ozueken A.S. (1998). Ethnic segregation in cities: new forms and explanations in a dynamic world. *Urban Studies*, 35, 1631-1656.

Vicari Haddock S. (2004). La citta' contemporanea. Bologna: il Mulino.

Chapter 8

How Does Immigration Impact on the Quality of Life in a Small Town?

James J. Potter, Rodrigo Cantarero, X. Winston Yan,
Steven Larrick, Heather Keele and Blanca E. Ramirez

Introduction

Throughout the world, people are migrating from rural areas to cities to find jobs and to seek a better life. As a result, many cities and regions are experiencing population influx. Many small midwestern communities have been experiencing these rapid demographic changes. 'In the 1990s, Nebraska rural counties continue(d) to experience out-migration of the young, working-age population, with remaining residents tending to be older. However, the 1980s and current out-migrants are being replaced by in-migrants' (Austin, 1996). Lured by jobs in the food processing industry (mainly meatpacking), many towns and cities have been experiencing a significant increase of new residents, mostly Latino, over the past 15 years. These changes tend to bring renewed economic vigor but they also present a complex array of physical, social, psychological, and cultural challenges. 'As in the case of Nebraska, an increase in the meatpacking industry and American minority and foreign immigration has definitely impacted rural communities, in terms of housing and quality of life issues' (Potter et al., 1996); coordination of human services (IANR, 1995); Hispanic migrant laborer homelessness (Gaber and Cantarero, 1997); community development (Gouveia and Stull, 1997); and language issues (Gouveia and Rousseau, 1995), (Ramirez, 1998) The resulting rapid and often unmanaged growth puts a strain on communities as well as on individual households. The fiscal viability of communities may be threatened, degradation of the environment can be exacerbated, and, in general, the quality of life may be diminished for both long-time residents and newly arrived residents. At the family level, individuals may suffer the stress of overcrowding, marital discord, child abuse, teenage rebellion, crime and a general loss of community identity.

This chapter reports a study that investigated issues pertinent to the impact of population influx due to migration on small cities, towns and communities. There are two primary reasons that led the researchers of this project to believe that the impact on small cities and towns was quite different from that on large cities. First, unlike large cities that often have greater resources, small cities and communities have limited resources to cope with pressures of population influx on their housing, infrastructure and municipal services. Big cities can readily absorb large numbers of immigrants due to their economies of scale, available housing of greater variety and social and cultural networks. Secondly, while the social and cultural environments of large cities are often diverse, small cities tend to be more socially, culturally and ethnically homogeneous. When the new comers are from a different cultural background, their arrival inevitably impacts the social and cultural homogeneity of the community.

This is precisely what has been happening in the city of Crete, Nebraska. Crete is located in Saline County and is approximately a 30-minute drive from Lincoln, which is Nebraska's Capitol City. Similar to other Nebraska counties, Saline County's demographic change can mainly be attributed to the growth in one or two communities. According to the US Census, Saline County had a total population of 12,715 in 1990 and by the 2000 Census it had experienced an increase to 13,843. This is an additional 1,128 residents (an 8.9% change). During the same time period, the Hispanic/Latino population grew by 911 persons in the County (a 9.85% change). Saline County's population increase is primarily due to the increase in Crete's population. Crete has been experiencing growth in population and change in the racial/ethnic makeup of its residents. According to the US Census, Crete's total population in 1990 was 4,841 and increased to 6,028 by the year 2000. This demonstrates a 24.5% increase in the population by 1,187 persons.

Crete was founded in 1870, partially due to the creation of Crete Mills, a Lauhoff Grain Company that produces food grain products. In 1910 Douglas Manufacturing opened (manufactures voting equipment), in 1965 Allen Products Company was established (now Friskies Pet Care Company, manufactures all meat dog food), in 1975 Farmland Foods was established (processes pork for the national and international market). In 1984 the first immigrants began arriving in Crete, Farmland Foods employed merely 12-15 Vietnamese workers to begin. In the later 1980s more Vietnamese families came to Crete for resettlement, and in the early 1990s Hispanics began to move to Crete from other meatpacking communities. In the late 1990s Bosnians moved to Crete through resettlement programs. Between December of 1999 and December of 2000, there was a 25% jump in the production work force in Crete, in part due to Farmland's ongoing expansion and high starting pay. Currently only half of Farmland's work force lives in Crete, or nearby Wilber, the other half lives in Lincoln. Crete's major industries are Farmland, Friskies, and Crete Mills. The company with the greatest number of immigrants is Farmland. Currently, Farmland has approximately 1,400 total employees, including production and administration. Because of the ethnically diverse employment population at Farmland, thirteen (13) languages are spoken at the plant. The different racial/ethnic groups employed by Farmland are Hispanic (250-300); Vietnamese (200-250); Bosnian (50-60); Chinese; Korean; Laotian; Croatian; Russian; Iraqi; etc.[1]

Although, the migration of persons from various ethnicities may have presented challenges to Crete leaders and community residents, this has also given

[1] Information provided by Farmland representative.

opportunities to engage in more cultural diversity dialogues and events. According to Farmland officials, the company has a Communications Committee designated to create programs celebrating diversity. For example, in 1999, a soccer team including Hispanic and Vietnamese players received sponsorship from the company. Furthermore, the Hispanic community is becoming more established in Crete and they are opening up restaurants and grocery stores.

The demographic changes occurring in Crete are brought about by a number of factors: 1) The availability of employment opportunities at Farmland, a pork processing plant; 2) the resettlement of refugees through organizations located in Lincoln, and; 3) word-of-mouth invitations by current Crete residents to their friends, families and acquaintances. The wave of newly arrived residents adds new life to the city and to its community, especially when many of them came from a culture that is distinctively different from that of the current residents. On the other hand, a flood of new immigrants poses a variety of unexpected challenges to the city.

The most immediately experienced impact of the latest population influx was on the city's housing, especially its affordable housing. The new arrivals needed affordable housing to buy and to rent. The municipal services such as schools, public transportation systems and hospitals were also in great need for expansion in order to accommodate the population growth.

Impact on its previously rather homogenous social and cultural environment is also quite obvious. Like other small cities in many parts of the US, residents in Crete in general had been quite homogenous both culturally and ethnically until the recent population influx. The new population influx of people with distinctively different cultural and ethnic background exerted an impact on the existing cultural homogeneity, and triggered a process of transformation from cultural homogeneity to heterogeneity.

The Study Methods

In the fall of 2000, Steve Larrick, Community Development Coordinator for UNL's College of Architecture, received a request from the City of Crete to study housing and quality of life issues. The Quality of Life team met with representatives from the City of Crete, Farmland and Doane College. During the meeting, the research team described the 'Housing and the Quality of Life in Schuyler, Nebraska' study (Potter et al., 1996) which would be used as a model for a study in Crete. The representatives were in support of the research model. From 1995 through 1996 a group of architectural and planning researchers at the University of Nebraska-Lincoln conducted a study to learn about the impact of the population influx on the physical environment and quality of lives of Schuyler. This study had two main objectives. First, it wanted to find out residents' perceptions of the city and its change in the physical, social and cultural environment as a result of the population influx. Secondly, it intended to identify issues that had direct or indirect implications for city housing and planning. The

Crete study would help explore the impacts of a Nebraska small town environment on *newly arrived residents* (more than 15 years) and *long-time residents* (less than five years).

The Methodology Utilized in this Study can be Summarized in Four Phases

PHASE I: Establishing Trust The main goal of this phase was to build trust with city officials in order to gain their input on the housing and quality of life issues. In addition, this stage also became important because researchers familiarized themselves with general community resources as well as those serving newly arrived residents. An initial meeting of December 13, 2000 between UN-L researchers and Crete officials marked the first of several meetings where the information provided became the vehicle for coordinating focus groups, designing and testing the questionnaire, locating the target population, and hiring Doane college students as interviewers.

PHASE II: Identifying Issue The primary goal of this phase was to gather information from meetings, conduct a literature search, collect newspaper articles, utilize US Census data, and gather feedback from focus groups. The literature search (i.e., newspaper and journal articles) helped in providing a picture of the historical forces that have shaped the current condition in Crete, NE. Furthermore, the Focus Groups (at least one for each target group) became a means of better understanding community residents' perceptions of the housing and quality of life issues.

PHASE III: Surveying for Resident Perceptions In February, the Quality of Life Study Team began developing the survey instrument for Crete, while utilizing the Schuyler questionnaire as a baseline. In developing the questionnaire, the following were utilized to come up with a final draft: two focus groups with newly arrived residents, one focus group with long-time residents. Some of the issues identified by newly arrived residents were: integration, communication, understanding federal and local laws/rules/regulations, the impact of being undocumented on obtaining a driver's license/health insurance/auto insurance, lack of transportation, and need for diversified shopping areas. On February 27, 2001, researchers met with Crete Collaborating team to discuss the survey draft. Crete City officials gave feedback, and the final changes were made. The questionnaire was tested in English and Spanish and made available to Doane College students to conduct in the community. Doane students were trained as interviewers and other interviewers were utilized from another research project Dr. Rodrigo Cantarero was running.

The goal of the household studies was to provide an assessment of the housing conditions (as perceived by both the recently arrived residents and the long-time residents), the stress associated with the changes in the community and how their well being has been affected as a result of those changes. Participants of the Long-Time Resident Interviews were individuals who have lived in Crete prior to January 1986. Newly arrived residents were defined as any person who has lived in Crete since January 1996. No Crete residents who moved to Crete after January 1986 and before January 1996 would be interviewed. Besides meeting the resident requirement, potential participants had to meet the age requirement of currently being 20 years of age or older.

PHASE IV: Analyzing the Survey Results The data was analyzed using SPSS. The two groups were compared with regard to their responses to each survey question.

Survey Results and Discussion

When considering the results of the survey there are many ways to present the information. To facilitate understanding, the results have been organized into two sections. The first section addresses issues in which the perceptions of long-time and newly arrived residents are similar. The second section discusses those issues for which perceptions of long-time and newly arrived residents are significantly different.

Similarities in Perception

The following are areas where there were not significant differences between responses of long-time and newly arrived residents ($p < .05$).

Satisfaction of the Residents with their lives in Crete Do residents feel satisfied with their lives in Crete? Long-time and newly arrived residents are generally satisfied with housing issues within the context of their neighborhood and the city of Crete. Both the long-time and newly arrived residents agreed that they had no problem with rodents or bugs in their residences, the plumbing condition was generally rated good, as well as the condition of appliances. Long-time and newly arrived residents are satisfied with many other aspects of the residential neighborhood.

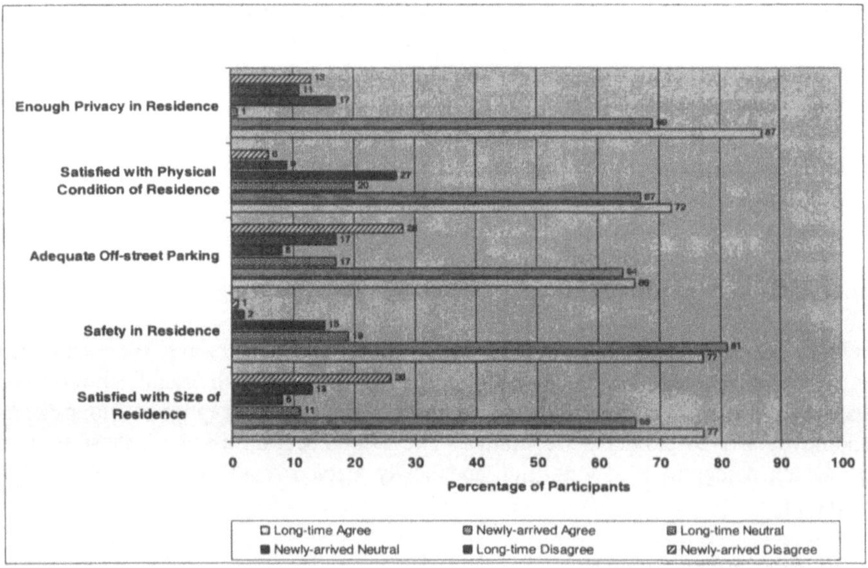

Figure 8.1 Perception similarities: Satisfaction with life in Crete

They agree that they have adequate off street parking, they are satisfied with the level of safety from being a victim of a crime while in their residence, they are satisfied with the overall physical condition of their residence, they are generally satisfied with the size of their residence, and they feel that they have sufficient privacy from neighbors.

Pressure on Public Municipal Services Do residents perceive there is a pressure on public municipal services? Long-time and newly arrived residents are satisfied with many public service issues within the city of Crete, as is evident in Figure 8.2.

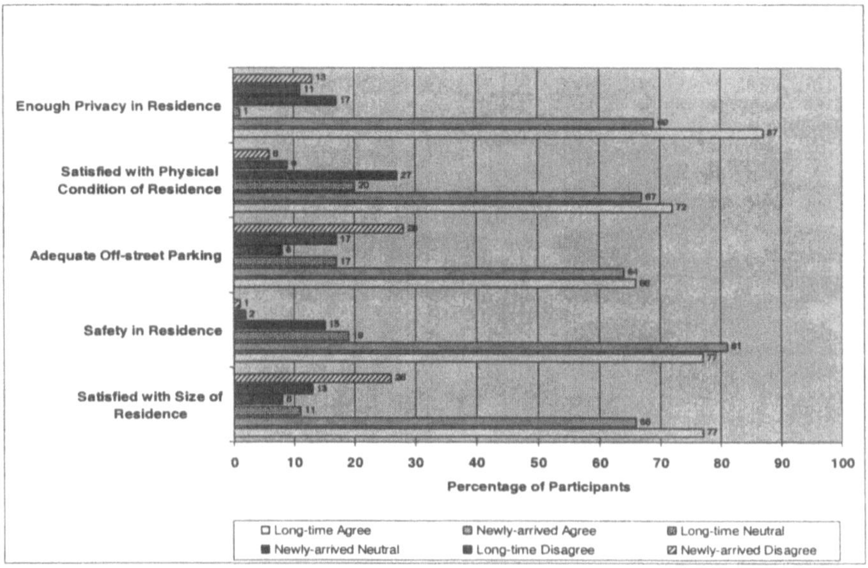

Figure 8.2 Perception similarities: Pressure on public municipal services

Both groups of residents are satisfied with the level of street maintenance and garbage collection in the neighborhood, and they rated the overall visual attractiveness of the neighborhood as satisfactory. Generally both groups rated the overall quality of air in the neighborhood as satisfactory, as well as the adequacy of public services. Long-time and newly arrived residents agree in their satisfaction with many aspects of service in the city of Crete. Both groups of residents are satisfied with the level of police protection and access to recreation services.

Stress of New Arrivals How did the new arrivals evaluate their new lives in Crete and what sources of stress did they experience most? As shown in Figure 8.3, long-time and newly arrived residents agree on many stress related concerns.

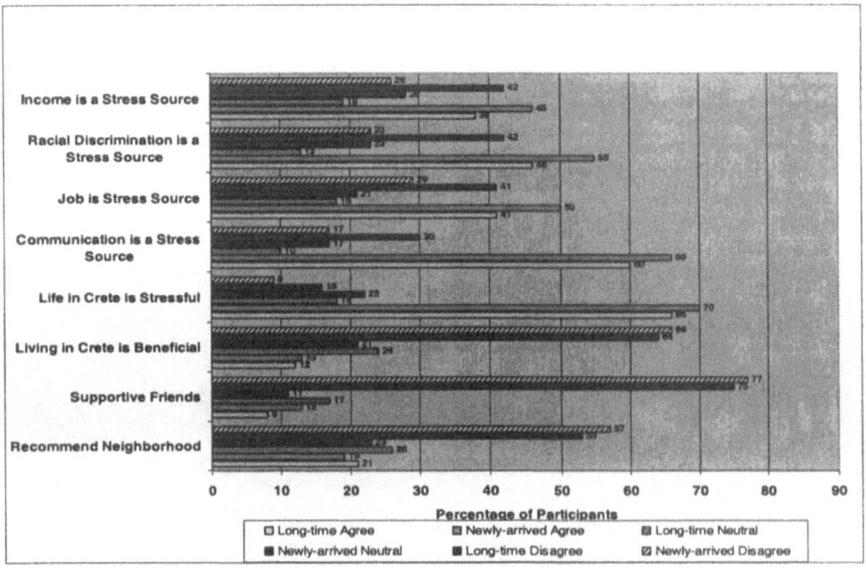

Figure 8.3 Perception similarities: Stress related concerns

Both groups agree that they can rely on friends for support in times of need, that living in Crete is beneficial for their family, and that they would recommend their immediate neighborhood to a friend. Both groups disagree with the statements life in Crete is very stressful and an inability to communicate with others is stressful for themselves. The subject of income as a source of stress garnered mixed perceptions among the groups, but generally neither group feels that income is a source of stress for them, with 45.7% of long-time residents and 38.5% of newly arrived residents disagreeing that income is a source of stress. In opposition to these findings, 42.3% of the newly arrived residents agreed that their level of income was a source of stress, while only 26.1% of long-time residents agreed their income was a source of stress. In general half of all residents disagree that racial discrimination is a source of stress. Both groups are split in their agreement concerning how their job (or lack of job) is a source of stress.

The Change in Neighborhoods: From Homogeneity to Heterogeneity Long-time and newly arrived residents agree on many aspects of the current housing condition in the city of Crete.

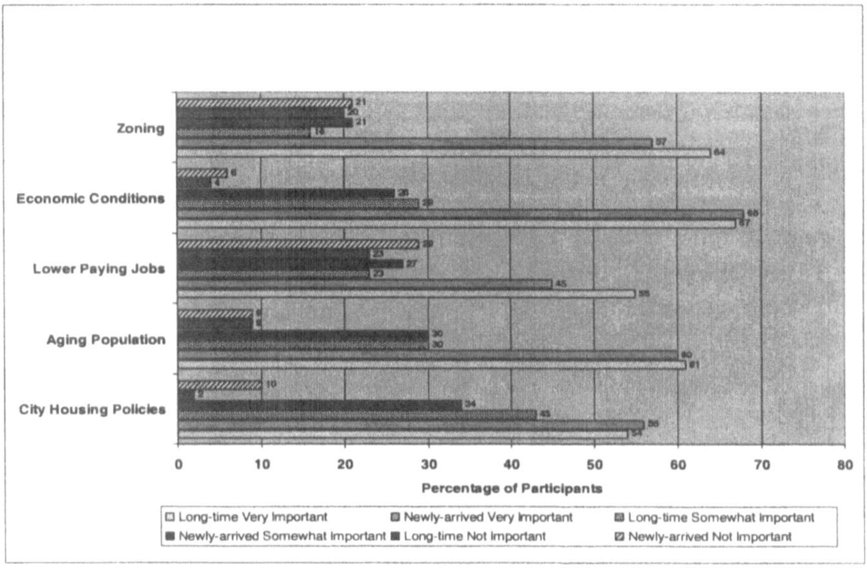

Figure 8.4 Perception similarities: Importance of factor in current housing condition

Both groups agree that the aging of resident population and the economic conditions, including lower paid jobs, are important factors contributing to the current housing conditions. They also agree that the city housing policies and zoning conditions are important factors contributing to the current housing conditions. Long-time and newly arrived residents agree on most social/cultural issues in the city of Crete. Both groups of residents are satisfied with the 'sense of community' in Crete, and with the cooperation among neighbors.

The Concern about Housing Priority In response to questions in the survey regarding priority of new housing construction in order to accommodate the population influx, it was discovered that long-time and newly arrived residents agree about several housing issues within the context of their neighborhood and the city of Crete. Both groups of residents rate the quality and maintenance of housing in their neighborhood as somewhat good. They agreed that the availability of rental assistance for families in Crete was somewhat good, noting that an average of 40% of the respondents fell into this category. There was a spread of ten-percent difference in this consensus though, with 35.9% of newly arrived residents rating it as somewhat good, while 45.2% of long-time residents thought it was somewhat good.

The Investment for New Housing Development Long-time and newly arrived residents have some similar priorities when it comes to improving housing conditions in Crete. Building more modest income houses is important to 72% of newly arrived residents and 79.5% of long-time residents. Both groups of residents are equally divided when it comes to rating the importance of building more upscale homes, on average 42.5% of long-time and newly arrived residents rated this as less important, while an average of 32.3% of both groups rated this as important, so they agree but do not form a strong consensus. Building more rental apartments is important to an average of 75.3% of both groups however they vary in their degree of agreement. Over 80% of newly arrived residents view this as important while 69.6% of long-time residents view it as important.

Differences in Perception

The following are areas where there were significant differences between responses of long-time and newly arrived residents ($p < .05$).

Satisfaction of the Residents with their lives in Crete Long-time and newly arrived residents are generally satisfied with housing issues within the context of their neighborhood and the city of Crete; however, the degree of privacy they feel they have from others in their residence differs. More than 86% of long-time residents agree that they have enough privacy from others in their residence, while only 69% of newly arrived residents agree.

The Pressure on Public Municipal Services Do residents differ in their perception of pressure on public municipal services? Long-time and newly arrived residents only have one response to physical issues in which the two groups had a difference of perception. Newly arrived residents were typically satisfied with the traffic that goes through their neighborhood, while long-time residents were neutral on this topic.

The Stress of New Arrivals How did the new arrivals evaluate their new lives in Crete differently than long-time residents? Figure 8.5 illustrates the differences between the groups.

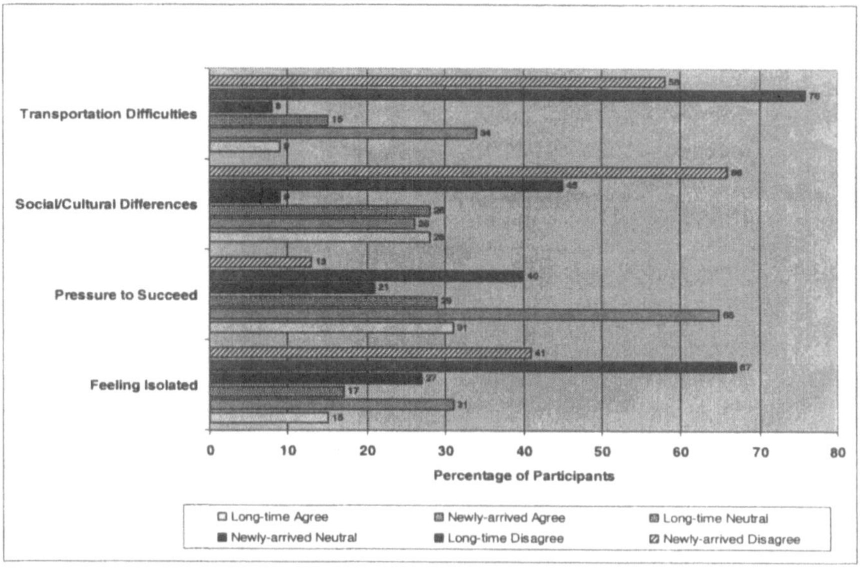

Figure 8.5 Perception similarities: stress related concerns

Long-time residents disagree by 67.4% that feeling isolated is a source of stress, while only 41.2% of newly arrived residents disagree. Even more significant, 31.4% of newly arrived residents agree feeling isolated is a source of stress for them. Long-time residents disagree that they feel pressure to do better, advance or succeed, and newly arrived residents do agree they feel pressure to do better. Social or cultural differences of people in the community are a source of stress for 66% of newly arrived residents, while 44.7% of long-time residents disagree that the social or cultural differences of people in the community are a source of stress. This is a large difference in their perception, however 27.7% of long-time residents remained neutral on this topic. Transportation is a source of stress for 34.0% of newly arrived residents, and 76.1% of long-time residents do not perceive transportation to be a source of stress. However, 58.0% of newly arrived residents disagreed that lack of transportation is a source of stress.

The Change in Neighborhoods: From Homogeneity to Heterogeneity The majority of both long-time and newly arrived residents feel that the population increase is an important factor that contributes to the current housing condition, as seen in Figure 8.6.

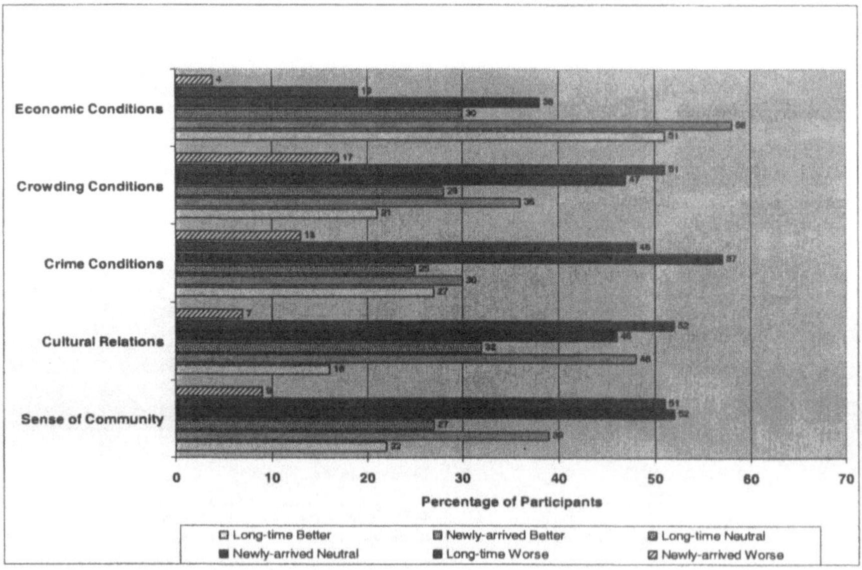

Figure 8.6 Perception similarities: changes in Crete

Where they differ is the level of their agreement, with 87.0% of long-time residents and 66.7% of newly arrived residents agreeing on the importance of this factor. Conversely, 25.5% of newly arrived residents were neutral in response to this, and 7.8% of newly arrived residents felt that this was less important, and no long-time residents viewed this factor as less important. An increase of industrial activity in the area is an important factor contributing to the current housing condition, with an amazing 91.5% of long-time residents perceiving this as important, and only 72.0% of the newly arrived residents agreeing.

Long-time and newly arrived residents have different perceptions concerning the change taking place in their neighborhoods. Long-time residents feel that the sense of community has gotten worse while newly arrived residents generally feel that it has gotten better. Long-time residents also feel that cultural relations have gotten worse, and newly arrived residents feel that they have gotten better. Long-time residents feel that crime conditions have gotten worse, and newly arrived residents feel that they have gotten better. Long-time residents feel that crowding conditions have gotten worse, and newly arrived residents feel that they have gotten better. The way the two groups view economic conditions also varies, with 57.7% of newly arrived residents, and 51.1% of long-time residents feeling they have improved, however 19.1% of long-time residents feel that it has gotten worse, as compared to 3.8% of newly arrived residents.

The Concern about Housing Priority In response to questions in the survey regarding priority of new housing construction in order to accommodate the population influx, it was discovered that long-time and newly arrived residents agreed on many points. It is interesting to view this agreement of perception in the context of their differing housing circumstances, with the majority of newly arrived residents renting and the majority of long-time residents as homeowners.

The Investment for New Housing Development Long-time and newly arrived residents have some differences in priorities when it comes to improving housing conditions in Crete. Building more housing for the elderly is important to 90.7% of long-time residents, while only 53.1% of newly arrived residents feel it is important.

Building more trailer parks is important to 38.8% of newly arrived residents, and is less important to 75.6% of long-time residents. However, 46.9% of newly arrived residents also feel it is less important.

Discussion

More than anything, the study discovered that the impact of the population influx on the city was felt in many aspects of its physical and cultural environment. And the impact is indeed significant. The population influx has exerted a great pressure on housing availability in the city, especially affordable housing. It has also changed the city's social and cultural landscape. It is evident from the survey results that long-time and newly arrived residents have both, similarities and differences, in their perceptions regarding Crete's quality of life.

Overall, Crete residents enjoy a good quality of life. In the majority of the survey results, residents expressed an appreciation for the good quality of life in Crete, and said that there is a 'sense of community' in Crete, and cooperation among neighbors. Residents said they could rely on friends for support in times of need, that Crete is beneficial for their family, and that they would recommend their immediate neighborhood to a friend. Residents feel safe from crime while in their residence, and there is a good level of police protection. They also feel that there is adequate off street parking, the level of street maintenance in the neighborhood, garbage collection, and the overall visual attractiveness of the neighborhood are good, as well as access to recreational services.

Additional efforts are needed to better meet the housing needs of Crete. Both long-time and newly arrived residents agree on factors contributing to the current housing conditions, as well as on the housing priorities of the community. Residents view the population, industrial activity, city housing policies, an aging population, lower paying jobs, economic conditions, and zoning as the major factors influencing current housing conditions. The residents agreed that housing priorities for the community include modest income housing, rental apartments, and elderly housing.

Long-time residents tend to feel conditions are worsening, while newly arrived residents tend to feel conditions are improving. It is important to notice the split in perception between the two groups of residents, and these observations can be found in the survey results. The major areas of concern include the 'sense of community', cultural relations, and crowding conditions.

Newly arrived residents tend to feel higher levels of stress than long-time residents do. In general, both groups do not feel that life in Crete is very stressful, however the amount of stress perceived is different. Newly arrived residents feel higher amounts of stress than long-time residents due to income levels, communication difficulties, feelings of isolation, their job, racial discrimination, and lack of transportation.

In Comparison to Schuyler

The findings from this study of the impact of the population influx in Crete can be compared to the findings of a previous study of Schuyler, Nebraska. There are slightly different conditions in Schuyler than in Crete. In Schuyler, most of the newly arrived residents were Latino, and of those an overwhelming majority was of the Catholic faith. Whereas in Crete, there was a much more diverse group of newly arrived residents, both culturally and religiously. However, both cities have experienced a major population influx due to meat-packing plants in their community, and both studies utilized the same base survey and methods. Analyzing their results in parallel creates a larger base of information from which to learn, and some conclusions can be made.

Quality of Life Issues In Crete and Schuyler the long-time and newly arrived residents agree that they feel safe from being a victim of crime while in their residence, they are satisfied with the overall physical condition of their residence, and feel they have adequate off street parking. They also agree that there is sufficient privacy from neighbors.

Stress Issues In Crete and Schuyler the long-time and newly arrived residents agree that living in their community is beneficial for their family. They also agree that they can rely on their friends for support, and that they would recommend their immediate neighborhood to a friend. In Crete and Schuyler the newly arrived residents feel a greater pressure to do better, advance or succeed than the long-time residents. The long-time and newly arrived residents also disagreed regarding racial discrimination. None of the groups viewed racial discrimination the same way.

Housing and Homogeneity Issues In Crete and Schuyler the long-time and newly arrived residents agree that there is a good 'sense of community', and they feel there is cooperation among neighbors. They also all agree that the quality and maintenance of housing in their neighborhood is good, that their community should have priorities to build more modest income houses and rental apartments. In Crete the two groups agree that the availability for rental assistance is somewhat good, while in Schuyler the two groups said it was poor. In Crete and Schuyler the long-time residents felt that the crime conditions and the crowding conditions have gotten worse, while the newly arrived residents feel they have gotten better.

Conclusion

At the end of each study the Quality of Life team offered various recommendations and possible actions for the community to take to improve the quality of community life. The Quality of Life team offered the following suggestions for the city of Crete. These suggestions were as follows: 1) Build on existing community initiatives that can make Crete's good quality of life even better; 2) Continue to explore ways to better provide for local housing needs in Crete; 3) Enhance community unity and cohesiveness by bridging the gap in perceptions between long-time and newly arrived residents; 4) Help make the transition easier and less stressful for newly arrived residents. It was suggested to the Crete community that they use their own expertise to come up with more recommendations and to pursue a plan of action that best fits the individual needs of the residents. What is down the line for this Quality of Life research? Currently the Quality of Life research team is pursuing the opportunity of carrying out a similar study in the community of Grand Island, Nebraska.

References

Austin, J. (1996). Migration in Nebraska Counties, 1980 to 2000. *Business in Nebraska* 52, no. 614 (October), 1-5.

Census 2000 data from www.census.gov

Gaber, S.L. and Cantarero, R. (1997). Hispanic Migrant Laborer Homelessness in Nebraska: Examining Agricultural Restructuring as One Path to Homelessness. *MARS/Social Thought and Research* 20(1-2), 55-72.

Gouveia, L. and Rousseau, M. (1995). Talk is Cheap: The Value of Language in the World Economy-Illustrations from the United States and Quebec. *Sociological Inquiry* 65(2), 156-180.

Gouveia, L. and Stull, D. (1997). *Latino Immigrants, Meatpacking, and Rural Communities: A Case Study of Lexington, Nebraska* Michigan State University: The Julian Samora Research Institute, unit of the Colleges of Social Sciences and Agriculture and Natural Resources. Research Report No. 26.

Potter, J., Cantarero, R., Yan, W., Larrick, S. and Ramírez, B. (1996). Residents' Perceptions of Housing and the Quality of Life in Schuyler, Nebraska. Lincoln: Final Report, University of Nebraska, College of Architecture.

Ramirez, B.E. (1998). The Influence of Micro and Macro Forces on Agriculture and Migration: Case Study of Hispanic Meatpackers in West Point, Thesis (M.C.R.P.), Lincoln: University of Nebraska.

University of Nebraska-Lincoln Cooperative Extension, IANR. 1995. *Collaboration for Community Problem-Solving: Lexington, Nebraska 1993-1995.* Lincoln: University of Nebraska – Lincoln Cooperative Extension, IANR.

Chapter 9

House Design as a Representation of Values and Lifestyles: The Meaning of Use of Domestic Space

Ritsuko Ozaki

Introduction

The design of 'home' has 'reflexive power', conveying our lifestyles and associated values and norms to others and back to ourselves. In sociology, social psychology and phenomenology, a number of studies have explored 'the meaning of home' (e.g. Gurney 1990, Perkins and Thorns 1998, Perkins and Thorns 2000, Saunders 1990, Saunders and Williams 1988, Somerville 1997). These 'meaning of home' studies began by investigating meanings in the typical middle-class household living in an owned single-family house, and have come to explore more diverse household types in such contexts as unemployment, retirement and single-parenthood (Després, 1991). More recent research covers a variety of home settings from private tenancy (Knight, 2001) to home working (Moore, 2001) and explores different experiences and meanings of 'home'.

These studies mainly look at the concept of 'home' and do not always deal with its physical aspects, such as internal layouts and features. However, there are meanings attached to design, too. In the practice of our everyday life, 'meanings' are inscribed in our *use of space*. Places have meanings, symbolize who belongs there and reflect images and activities (Zukin, 1995). Our home is the main locale of our daily activities. We spend most of our lives there; it is the place where familiar routines and social interactions take place (Saunders, 1990). How we use and 'appropriate' our domestic space is a symbolic practice and is a cultural phenomenon. This intentional use of space is 'cultivation' of space and creates the difference between 'house' and 'home' (Lawrence, 1995). This cultivation of space is indeed a process of making a house into a home (Perkins and Thorns, 2000). Therefore, the use of domestic space, and accordingly, the design of 'home' reflects our underlying values and norms (Clark, 1973; Mumford, 1970; Ozaki, 2001, 2002a); in other words, basic boundaries in domestic space are a metaphor reflecting taken-for-granted assumptions and ideologies (Jordanova, 1990). As Chaney (1996) puts it, design becomes more important than function as we use and appropriate objects and attach meanings to them.

In this context, this chapter considers physical aspects – i.e. the design – of 'home'. It first discusses the importance of analyzing the meaning of use of space in exploring the meaning of the design of 'home', and presents the relationship between 'lifestyles and associated values and norms' and 'the use and design of domestic space' in Britain from a historical perspective. Secondly, the chapter investigates what the design of contemporary British homes may mean through the analysis of the way in which people use their domestic space, reporting on the progress of an on-going survey with those who are on a 'shared ownership' scheme.

Rituals and Use of Domestic Space

In sociology, 'lifestyles' are seen as sets of shared values, practices and attitudes that make sense in particular contexts, such as patterns of social relations and consumption activities (Abercrombie et al., 1984). The concept of lifestyles explains what people do, why they do it, and what doing it means to them. Lifestyles reflect *how people use* what they have (Chaney, 1996); and different ways of using goods, places and times create bonds or distinctions among people (Featherstone, 1991). In other words, how we use domestic space in our daily lives expresses our lifestyles.

What we do in our everyday life regulates our ways of using space. 'Rituals' are an attempt to maintain a particular culture or a particular underlying assumption by enacting them visibly, and therefore a ritualized activity such as cooking is to be seen as a cultural expression (Douglas, 1966). Indeed, 'even when physical possibilities are numerous, the actual chores may be severely limited by the cultural matrix', such as social conventions and taboos (Rapoport, 1969: 47). Seen this way, how we use our living space to carry out rituals in our everyday life is a key to understanding the meanings that we attach to our home. As Lawrence (1990) puts it, if we are to understand meanings inscribed in the design of 'home', we need to analyze how our behaviors and activities are regulated by codes and norms. Rituals therefore have a central role in exploring use of domestic space.

Among the salient rituals traditionally conducted in British society are formality rituals – both to express social status of the family and to maintain formality within the household. The formality rituals have had a significant impact on the way domestic space is used – namely, the perception of 'front' and 'back' regions of the house.

Historical Perspectives on Formality Rituals in Britain

This section outlines changes in formality rituals in British society in the past to see how our ways of using space have expressed our lifestyles and associated values and norms.

Symbolic sphere vs. secular sphere

Houses in Britain have traditionally had a demarcation between 'front' and 'back' regions (rooms). This distinction between the front and the back is closely associated with the demarcation between the public, symbolic life and the private, secular life. In the nineteenth century, industrialization, promoting the view that one's home was a private arena away from the unpleasantness of the outside world, brought the segregation of living patterns, especially in middle-class households[1]. First, space for work disappeared from the living area; a house was physically separated from business. Then, inside the house, 'back-stage' functions, such as cooking, eating, washing and sleeping, began to be separated from polite, social activities. Within the domestic area, the upper- and middle-class house came to have special places for each activity, which were set aside for the specific purposes, like the front parlor, the dining room and the kitchen. Servants' quarters were located in the back premises, since they functioned as part of back-stage activities[2] (Williams, 1987).

By the end of the nineteenth century, this demarcation had reached the working-class household, which led to a widespread realignment in the use of the private domain in the working-class house. Skilled workers, such as artisans, began to live in two-storey terraced houses, which consisted of a front room, a back room and a scullery on the ground floor with bedrooms on the first floor. The front room was used for best occasions, the back room was for cooking and family eating, and the scullery was for washing activities (Daunton, 1983)[3]. Inevitably, there were strong aspirations for a higher-class lifestyle and upward social mobility among the working class (Harris, 1993). They placed a strong emphasis upon the front region of the house. People papered the walls, covered the floor and flagstones with oilcloth, laid carpets and put brass fenders and even a piano in the front room. They sought the separation of functions that the middle class had made in the earlier part of the century, as a way of showing respectability and to distinguish one's self and family from the lower classes (Daunton, 1983).

Social relations within the household

The demarcation between the front and the back did maintain formality relating to the male-female, adult-child and family-servant distinctions within the Victorian household. Middle-class women were educated to be subservient to men and were confined to the private sphere in the back region of the house. Men, on the other

---

[1] Middle class households had the strictest boundaries between private and public spheres, while the upper class still had estate management and legal or political duties in the homes, and the working class still spent their private time after work in streets or taverns (Davidoff and Hall 1987).

[2] This separation between families and servants is also a result of formal social relations within the household, as can be seen in the next section.

[3] In smaller terraces with two rooms on the ground floor, one would use the room in the front as the front parlour and the back room as the kitchen and the scullery (Daunton 1983).

hand, used the billiard room, the smoking room, the library or study and the dining room, which were situated in the front part of the house and were regarded as the 'male sphere'. Children were kept away from the front region, and the servants were confined to the back region (Burnett, 1986; Williams, 1987).

The decline in domestic service between 1890 and 1940, which resulted from the large demand for labor in offices and factories (Horn, 1975), had two effects on the middle-class household. One was the disappearance of servants' quarters (Saunders, 1990). The other was the change in women's roles[4]. The scarcity of domestic servants forced middle-class women to take over the servants' work and required them to preserve the same standards and routines (Ravetz, 1989); they had to play the roles both of a domestic servant and of the hostess (Goffman, 1959). The traditional role of women and the lowly status of domestic labor began to change (Saunders, 1990). Consequently, middle-class families developed less formal relationships between husbands and wives, and parents and children, which led to a decline in formality within the household (Burnett, 1986).

These changes within the household led to changes in the use of domestic space. Formerly, there had been a strict demarcation between the kitchen for cooking activities, which was a female domain, and the dining room for eating formal meals, which was a male domain. The new conjugal roles brought a new way of eating, that is, people ate in the place where the meal had been prepared (Burnett, 1986; Saunders, 1990). While the joint conjugal lifestyle made an appearance in the middle-class household, the working class, who started to utilize the demarcation between the front and the back, were more likely to retain conventional segregation of the lives of the husband and wife (Jansson, 1995).

The above historical account shows the link between lifestyles and use of space. The segregation of use of front and back regions expresses an acute status consciousness and represents formal relationships among household members. A more integrated way of using domestic space reflects less formal relations between men and women and between adults and children. These meanings of use of space then indicate the meanings of the design of 'home': a segregated front room reflects social status of the family and male authority within the household, and integrated rooms (e.g. a kitchen-diner) mean more democratic relations among household members. Clearly, the analysis of the meaning of use of space offers a useful perspective in exploring the meaning of the design of 'home' (see Ozaki, 2003, for a more detailed discussion on the use of front and back regions in the British house and its expression of changing values and lifestyles).

Empirical Study of Use of Space

In that case, how do we use our domestic space nowadays, and what do our ways of using space represent in terms of our lifestyles, and consequently what meanings

---

[4] Apart from the decline in domestic service, the reduction in family size and birth rates contributed to the emergence of smaller households and to the broader emancipation of women, together with new legal frameworks for women (Harris 1993; Stevenson 1984).

are attached to the design of 'home'? In order to answer these questions, a questionnaire survey was conducted in co-operation with a London-based housing association. The sample in this particular survey is 'shared owners'. The association's shared ownership scheme aims to provide low-cost, affordable houses for first-time buyers who want to buy in the open market but cannot afford to do so[5]. Shared owners are under-surveyed and the association is eager to understand what their preferences in internal layout are[6]. In April 2002, questionnaires were sent out to 113 home-purchasers in London and the South East of England who bought a house or a flat between April 2000 and March 2002. Forty-four questionnaires were returned (a 38.9 per cent response rate).

This survey focused on the configuration of the kitchen, dining and living room area, and addressed the following questions:

- Where do respondents eat their main meals, and why?
- How would they change the layout, if it is not their preferred way (the way they would like it)?
- How and where do they eat on special occasions or with guests, and why?
- How would they change the layout, if it is not their preferred way (the way they would like it)?
- Are there any other activities that they (or their family) undertake in these rooms?

Factual information (e.g. demographic data) was also requested. Open answers and comments were coded, and all the answers were subjected to statistical tests.

Respondents' profile

Respondents are first-time buyers aged between 25 and 45. Of the respondents, 64 per cent are aged between 25 and 35 and the rest are aged between 36 and 45. Some 64 per cent have no dependent children living in the home[7], 71 per cent are female, 73 per cent are white British with a quarter being African/Caribbean black or Asian (and one person declined to answer this question). With regard to the current room configuration, over half the respondents have a lounge-diner, one-

---

[5] Purchasers can buy an initial share (generally around 50 per cent of the value of the property) and pay a subsidised rent on the remaining value of the property. This offers shared-owners the opportunity to eventually own their home.

[6] The results have implications for the design of shared owners' housing. It is important for the UK housing associations to understand what attracts their customers, as shared owners are normally different from tenants in terms of occupations, household composition and ethnic background.

[7] The housing association has now developed a new questionnaire which asks the age of children. This way, more detailed analysis of the effect of children in the home (and their age) can be made in the future. Indeed, in their study of Italian homes, Giuliani et al. (1993) found that the age, as well as the number and gender, of children have significant influences on the furnishing and use of rooms.

third have a kitchen-diner, and the rest have either separate rooms or an all-in-one living room (Table 9.1).

Table 9.1 Respondents' profile

Categories	*Sub-categories*	*Frequencies*	*%*
Age	25-35	28	63.6
	36-45	16	36.4
Children in the home	Yes	16	36.4
	No	28	63.6
Gender	Male	13	29.5
	Female	31	70.5
Ethnic origin	White British	32	72.7
	Others (Black / Asian)	11	25.0
	Refused	1	2.3
Current room configuration	Lounge-diner and separate kitchen	24	55.8
	Kitchen-diner and separate lounge	14	32.6
	Separate kitchen, dining room and lounge	3	7.0
	All-in-one style living room	2	4.7

Responses

Where main meals are eaten seems to be dependent on the current room configuration, as shown by the fact that more than one-third of the respondents say that there are no alternatives (Table 9.2). Nonetheless, there are two contrasting attitudes towards eating activities – one is to see eating main meals as an occasion, and the other considers it to be casual and values convenience more highly.

Table 9.2 Reasons for using a particular room/space for eating main meals (coded answers)

- More casual, friendly way of eating is preferable – in the living room (26.8%)
- Sitting at the table is good/important (for children to learn manners) – at the dining table in the kitchen, living room or dining room (17.1%)
- Convenience is more valued – in the kitchen (12.2%)
- Meals are important family occasions – in a separate dining space (4.9%)
- Other rooms are too small, there are no alternatives (39.0%)

Similarly, the current room configuration is an important factor in determining where to eat with guests. Nevertheless, there are four different perceptions about how they eat when they entertain guests, with each perception placing importance on different values: homeliness/comfort, formality, convenience and sociability/inclusiveness (Table 9.3).

Table 9.3 Reasons for using a particular room/space for eating with guests (coded answers)

- More homely and comfortable atmosphere is preferable – in the living room (15.0%)
- Formal eating is important – in a separate dining space (13.5%)
- Convenience is more valued – in the kitchen (10.0%)
- More sociable and inclusive way of eating is favored – in the kitchen or in the dining area of the living room (2.5%)

Respondents' preferred ways of eating main meals and eating with guests can be divided into three categories: eating casually in the kitchen, eating formally in a separate dining room and eating more sociably in an integrated open room (Table 9.4).

Table 9.4 Preferred ways of eating main meals and eating with guests (coded answers)

Answers	*Eating main meals*	*Eating with guests*
Would make the kitchen larger and eat there casually	16 (42.1%)	18 (47.4%)
Have a separate dining room to eat formally	12 (31.6%)	11 (28.9%)
Integrate the kitchen into the living room and create an open space to eat more sociably	3 (7.9%)	3 (7.9%)
Happy as it is (with a kitchen-diner)	7 (18.4%)	6 (15.8%)
Total answers	38 (100.0%)	38 (100.0%)

The results show different perceptions about, and use of, space. How, then, are these perceptions associated with the respondents' demographic characteristics?

Associations with demographic factors

Demographic factors related to age, gender, ethnic origins and whether or not there are dependent children in the home (Table 9.5). With regard to 'use of space' factors, the following four issues were considered: reasons for using particular space for eating main meals and for eating with guests, and preferred ways of eating main meals and eating with guests.

Table 9.5 Predictor variables used to assess demographic factors

Independent variable	*Definition*
Age	1 = 25-35; 2 = 36-45
Children in the home	1 = Yes; 2 = No
Gender	1 = Male; 2 = Female
Ethnic origin	1 = White British; 2 = Non-white British (of other ethnic origins)

The results of Chi-square tests presented three significant associations between demographic factors and use of space. 'Preferred ways of eating main meals' and 'preferred ways of eating with guests' are associated with age ($p<.01$ and $p<.05$ respectively). Younger respondents are more likely to use a kitchen-diner to eat casually (including with guests) than older respondents. 'Reasons why they eat their main meals where they do' is associated with whether or not there are children in the home ($p<.05$). People without children like convenient and casual ways of eating. However, there are many cells with expected frequencies less than 5, and therefore the findings are not very reliable.

Since reliable significant relationships between demographic factors and use of space could not be obtained, discriminant analysis was applied for further analyses. However, only one statistically significant (and one nearly significant) result was obtained. This appears to be the effect of small cell groups; thus, more reliable results could be obtained with a larger sample in the future. Accordingly, this paper presents what the outcomes *indicate* as a stepping stone to the future survey.

First, with regard to reasons for using particular space for eating main meals, one function was estimated, with two variables (the importance of sitting at table/an important family occasion, and convenience/informality) and four predictors (age, gender, ethnicity and where or not there are children). In this analysis, the common response that there is no alternative to the present use was dropped from consideration. This function seems to imply that 'convenience/informality and the negative of 'good to sit at table/family occasion' are associated with 'no children' (Table 9.6).

Table 9.6 Discriminant analysis: Reasons for using particular space for eating main meals

Structure Matrix	
	Function 1
Children	**0.866**
Ethnicity	-0.431
Gender	-0.052
Age	-0.036
Functions at Group Centroids	
	Function 1
Good to sit at table/ important family occasions	**-0.765**
Convenient/ casual/ informal	0.430

*No overall significance (Wilks' Lambda, sig. = 0.169).

*Equality of variances as required for the validity of our tests of significance can be accepted (Box's M, sig. = 0.656).

*The classifications of original grouped cases are good (72.0% correctly classified).

Secondly, as for the reasons for using particular space for eating with guests, three functions were estimated, with four variables (sociability/inclusiveness, formality, convenience and homeliness/comfort) and the same four predictors. Again, responses citing 'no alternatives' were excluded from this particular analysis.

Some 61.6 per cent of the total variance is explained by Function 1, 35.6 per cent by Function 2 and 2.8 per cent by Function 3. Function 1 seemingly implies that 'homeliness/comfort' and the negatives of 'sociability/inclusiveness' and 'convenience' are associated with the 'older age group'. Function 2 appears to imply that 'convenience' and the negative of 'formality' are associated with the combination of 'older age group' and 'men'. Hence, the link between 'older age' and the negative of 'convenience' found in Function 1 mainly applies if the former consists of women (who make up the majority of all respondents); it is not the case for older men. Function 3 seems to imply an association between 'sociability/inclusiveness' and the combination of 'women' and 'no children' (Table 9.7).

Table 9.7 Discriminant analysis: Reasons for using particular space for eating with guests

Structure Matrix			
	Function 1	Function 2	Function 3
Age	**0.641**	**0.569**	0.470
Gender	0.153	**-0.676**	**0.668**
Children	-0.212	0.287	**0.934**
Ethnicity	0.152	0.116	-0.391
Functions at Group Centroids			
	Function 1	Function 2	Function 3
Sociable/inclusive	**-1.353**	-0.351	**0.674**
Formal	-0.293	**-0.892**	-0.117
Convenient	**-0.929**	**0.869**	-0.103
Homely/comfort-able	**1.089**	0.222	-

*No overall significance (Wilks' Lambda, sig. = 0.312).
*No Box's test of equality of covariance was performed because of the limited number of certain responses.
*The classifications of original grouped cases are good (62.5% correctly classified).

Thirdly, concerning the preferred ways of eating main meals, two functions were estimated, with three variables (making the kitchen into a large kitchen-diner to eat casually, having a separate dining room to eat formally, and integrating the kitchen into the living room to eat socially) and the same three predictors. All observations are included. Function 1 explains 76.2 per cent of the variance and Function 2 the other 23.8 per cent. Function 1 implies that 'an integrated open room', 'a separate dining room' and the negative of 'a kitchen-diner' are associated with the 'older age group'; and Function 2 suggests an association between 'an integrated open room' and 'no children' (Table 9.8).

Table 9.8 Discriminant analysis: Preferred layouts and ways of eating main meals

Structured Matrix		
	Function 1	Function 2
Age	**0.869**	0.397
Gender	0.221	0.172
Children	-0.424	**0.891**
Ethnicity	-0.134	-0.250

Lastly, regarding the preferred ways of eating with guests, two functions were estimated, with the same three variables and two predictors as above. Again, all observations are included in this analysis. Some 88.5 per cent of the variance is explained by Function 1 and 11.5 per cent by Function 2. Function 1 implies that 'an integrated open room' is associated with the 'older age group'. Function 2 does not suggest any significant associations (Table 9.9).

Table 9.9 Discriminant analysis: Preferred layouts and ways of eating with guests

Structured Matrix		
	Function 1	Function 2
Age	**0.642**	**0.606**
Children	0.419	**-0.767**
Gender	0.208	0.512
Ethnicity	-0.250	0.458
Functions at Group Centroids		
	Function 1	Function 2
Kitchen-diner, casual	-	-0.176
Separate dining room, formal	-0.434	0.369
Integrated open room, sociable	**2.221**	0.179

*Moderately high overall correlation (Wilks' Lambda, sig. = 0.056).
*Equality of variances can be accepted (Box's M, sig. = 0.895).
*The classifications of original grouped cases are quite good (56.8% correctly classified).

To sum up, households without children value convenience and casualness and would like to use an integrated open room for eating main meals. They do not regard their main meals as a family occasion and do not think it is good or important to sit at table. Women with no children value sociability and inclusiveness when they eat with guests, while men aged between 36 and 45 appreciate convenience, not formality. In general, older respondents like a homely or comfortable atmosphere, rather than sociability and convenience, when entertaining guests. They would prefer to eat their main meals either in an integrated open room or in a separate dining room, not in a kitchen-diner. They would also prefer to eat in an integrated open room with guests.

Discussion

These outcomes present three issues. First, results of both Chi-square test and discriminant analysis seemingly suggest some generational difference in attitudes towards eating activities, although respondents are relatively young people aged between 25 and 45 and the age gap within the sample is not large. Older respondents are more in favor of a homely/comfortable atmosphere, are not concerned about convenience and sociability, and would not prefer to use a kitchen-diner. This may be because they associate proper eating space of the dining room or spaciousness (but not necessarily sociability) of an integrated room with a homely and comfortable feel and do not value the convenience of a kitchen-diner. Younger respondents, on the other hand, are keener to have casual/convenient ways of eating and would prefer to use a kitchen-diner for eating main meals and entertaining guests.

Indeed, younger generations tend to eat more savory snacks or take-aways. Guests are often invited into the kitchen to have dinner (The Dietary and Nutritional Survey of British Adults 1990). The formal manner of having meals in a specially kept space, which has long been a ritual maintaining formality, has declined in importance. People have become more casual about eating meals and receiving guests, although there are still those who would only consider entertaining guests in the front room or who regard the dining room as important or appropriate for particular occasions (Ozaki, 2002b). Thus, it could be said that social trends towards younger generations being more informal in their lifestyles are well reflected by their ways of using space.

Secondly, respondents without children in the sample do not seem to have much respect for the traditional formality ritual. To them, meals are not a family occasion and sitting at table is not a particularly good thing. Instead, they attach importance to convenience and casualness. In reality, the formal family dinner has become a less important practice. As seen above, people increasingly eat ready-made foods at times that are convenient for them (The Dietary and Nutritional Survey of British Adults 1990). Thus, respondents' casual eating activities would appear to reflect such trends.

Lastly, our female respondents (with no children) seem to value inclusiveness and male respondents (aged between 36 and 45) appear to appreciate convenience and less formality in their use of space. This may be a reflection of a decline in the traditional formal gender relations; indeed, the kitchen integrated in the dining or living room is a sign of less segregation in the domestic division of labor (Hasell and Peatross, 1990; Zukin, 1982). Many women in Britain have actually improved their social positions: female full-time enrolments in higher education, labor participation, gross earnings and job-related training have all been on the increase (Social Trends, 1995). Also, the number of people who think men and women should share domestic tasks has risen in the last few decades, though women still tend to do more housework even when in full-time employment (British Social Attitudes, 1992). The more hours the female partner is in employment, the less conventional is the domestic division of labor (Pahl, 1984)[8]. But the degree of segregation of conjugal roles also depends on the family's lifecycle stages; some women remain at home while children are young (British Social Attitudes, 1995). Thus, the outcomes of this case study seem to express more equal gender relations, together with the issue relating to the effect of lifecycle stages.

To conclude, lifestyles and social relations of contemporary British people have become less formal, and, accordingly, formality rituals are on the decline, although there are lifecycle-stage and generational differences. These trends appear to be expressed by the ways in which our young respondents use space. Seen this way, in the present sample of 44 shared-owners aged between 25 and 45, it could be said that a kitchen-diner represents casualness and convenience (although when it first made appearance it mainly meant joint conjugal roles), that a separate dining room means formality (as before), and that an integrated room expresses not only sociability and inclusiveness, but also a homely and comfortable atmosphere (which is an unanticipated meaning). It is my belief that when a larger sample is taken, more significant results about the relationship between lifestyles and use of space will be obtained, and consequently more accurate meanings attached to the design of 'home' will be found.

Concluding Remarks

Use of space is an indicative way of investigating the meaning of the design of 'home'. The historical account, and the case study, although it could not present statistically significant results, have shown the relationship between our lifestyles and ways of using domestic space. This has helped us to explore the meanings attached to specific designs. Also, the results of the case study indicate that our respondents' preferred ways of using space for eating activities reflect their increasingly less formal lifestyles and social relations, showing different meanings of the kitchen, dining and living room area according to age, household composition and gender.

Needless to say, meanings attached to space do change. Our home keeps serving new functions as we attach new meanings to it, according to the changes in our lifestyles. For example, 'home' has become a workplace for some people. The progress of telecommunications and widespread use of the personal computer, together with mid-life redundancies, have increased the number of self-employed people who work from home (Perkins and Thorns, 1998). In our sample, 13 respondents say that they use the living room or lounge as a workplace. The

[8] Decision-making within the household is closely linked with the wife's employment status and contribution to the household income; and women who make no financial contribution to the household are still likely to be subject to male dominance in decision-making (Pahl 1989). Migration to suburban estates has promoted joint conjugal roles in marriage with more positive attitudes towards equal partnership in the traditional working-class household where a couple would have segregated conjugal relationships (Jansson 1995).

implication of changes in lifestyles is that the pattern of use of space has been changing.

This study has focused on shared owners and their preferences, but further work could investigate different market brackets and examine sub-cultural differences (e.g. shared owners vs. tenants), in order to gain a deeper understanding of the meaning of different internal layout and use of domestic space of the English house. Further work should also be conducted on a larger sample, with comparisons between households with children of different ages, so that the weaknesses of this study can be addressed and the effect of different lifecycle stages of respondents explored.

Our lifestyles have a significant effect on how we define the role of rooms, what activities we carry out there, and consequently what room configuration we would prefer. The analysis of the meaning of use of domestic space thus provides a clue to advancing our understanding of the meaning of the design of 'home'.

Acknowledgement

I would like to thank the housing association who has conducted this survey and the questionnaire respondents. I also wish to thank Nick von Tunzelmann for his help with statistics and John Rees Lewis for his comments on earlier versions of this chapter.

References

Abercrombie, N., Hill, S. and Turner, B. S. (1984). *The Penguin Dictionary of Sociology*, 2nd ed., London: Penguin Books.

Burnett, J. (1986). *A Social History of Housing 1815-1985*, 2nd ed., London: Routledge.

Chaney, D. (1996). *Lifestyles*. London: Routledge.

Clark, E. (1973). Order in the Atoni house. In R. Needham (ed.) *Right and Left: Essays in Dual Symbolic Classification*. London: The University of Chicago Press.

Central Statistical Office (CSO). *Social Trends*, 1995.

Daunton, M. J. (1983). *House and Home in the Victorian City: Working-Class Housing 1850-1914*. London: Edward Arnold.

Davidoff, L. and Hall, C. (1987). *Family Fortunes: Men and Women of the English Middle Class 1780-1850*, London: Hutchinson.

Després, C. (1991). The meaning of home: literature review and directions for future research and theoretical development. *Journal of Architecture and Planning Research*, 8(2), 96-115.

Douglas, M. (1966). *Purity and Danger: An Analysis of the Concepts of Pollution and Taboo*. London: Routledge and Kegan Paul.

Featherstone, M. (1991). Consumer Culture and Postmodernism. London: Sage.

Giuliani, M.V., Bove, G. and Rullo, G (1993). The spatial organization of the domestic interior: the Italian home. In E.G. Arias (ed.) *The Meaning and Use of Housing: International Perspectives, Approaches and their Applications*. Aldershot: Avebury.

Goffman, E. (1959). *The Presentation of Self in Everyday Life*. New York: Doubleday.

Gurney, C. (1990). The meaning of home in the decade of owner occupation: towards an experiential perspective, SAUS, University of Bristol, Working Paper 88.
Harris, J. (1993). Private Lives, Public Spirit: Britain 1870-1914. London: Penguin Books.
Hasell, M. J. and Peatross, F. D. (1990). Exploring connections between women's changing roles and house forms. *Environment and Behavior*, 22(1), pp. 3-26.
Horn, P. (1975). *The Rise and Fall of the Victorian Servants*. Dublin: Gill and Macmillan.
Jansson, S. (1995). Food Practices and Division of Domestic Labour: A comparison between British and Swedish households. *The Sociological Review*, 43(3), 462-477.
Jordanova, L. (1990). *Sexual Visions*. London: Harvester Wheatsheaf.
Knight, D. (2001). Are we home yet?, paper presented at the HSA autumn conference, Cardiff, 4-5 September.
Lawrence, R. J. (1990). Public collective and private space: a study of urban housing in Switzerland. In S. Kent (ed.) *Domestic Architecture and the Use of Space: An Interdisciplinary Cross-Cultural Study*. Cambridge: Cambridge University Press.
Lawrence, R. J. (1995). Deciphering home: an integrative historical perspective. In D. Benjamin (ed.) *The Home = Words, Interpretations, Meanings, and Environments*. Aldershot: Avebury.
Moore, J. (2001). Toil or retreat: the experience of home for homeworkers, paper presented at the HSA autumn conference, Cardiff, 4-5 September.
Mumford, L. (1970). *The Culture of Cities, 1970* ed.. Westport: Greenwood Press.
Office of Population Censuses and Surveys (OPCS) (1990) *The Dietary and Nutritional Survey of British Adults*, London: HMSO.
Ozaki, R. (2001). Society and housing form: home-centredness in England vs. family-centredness in Japan. *Journal of Historical Sociology*. 14(3), 337-357.
Ozaki, R. (2002a). Housing as a cultural representation: privatised living and privacy in England and Japan. *Housing Studies*. 17(2), 209-227.
Ozaki, R. (2002b). Mind the gap: Customers' perceptions and the gaps between what people want and what they are offered, the Joseph Rowntree Foundation (ed.) *Consumer Experiences in UK Housing*, The Joseph Rowntree Foundation, York.
Ozaki, R. (2003). The 'front' and 'back' regions of the English house: changing values and lifestyles. *Journal of Housing and the Built Environment*, 18(2), forthcoming.
Pahl, J. (1989). *Money and Marriage*, Basingstoke: Macmillan.
Pahl, R. E. (1984). *Division of Labour*, Oxford: Blackwell.
Perkins, H. C. and Thorns, D. C. (1998). House and home: looking in on New Zealanders' culture, sense of identity and sense of place, paper presented at RG43: Housing and the Built Environment, International Sociological Association World Congress, Montreal, 25 July-5 August.
Perkins, H. C. and Thorns, D. C. (2000). Making a home: housing, lifestyle and social interaction, paper presented at the ENHR 2000 conference, Gävle, Sweden, 26-30 June 2000.
Rapoport, A. (1969). House Form and Culture. Englewood Cliffs, N. J.: Prentice-Hall.
Ravetz, A. (1989). A View from the Interior. In J. Attfield, and P. Kirkham (eds.). *A View from the Interior: Feminism, Women and Design*. London: The Women's Press.
Saunders, P. (1990). *A Nation of Home Owners*. London: Unwin Hyman.
Saunders, P. and Williams, P. (1988). The meaning of 'home' in contemporary English culture, *Housing Studies*. 4(3), 177-192.
Stevenson, J. (1984). *British Society 1914-45*. London: Penguin Books.
Social and Community Planning Research *British Social Attitudes*, 1992, 1995.
Somerville, P. (1997). The social construction of home. *Journal of Architectural and Planning Research*. 14(3), 226-245.

Williams, P. (1987). Constituting class and gender: a social history of the home, 1700-1901. In N. Thrift and P. Williams (eds.). *Class and Space: The Making of Urban Society*.London: Routledge and Kegan Paul.
Zukin, S. (1982). *Loft Living: Culture and Capital in Urban Change*. Baltimore: The Johns Hopkins University Press.
Zukin, S. (1995). *The Cultures and Cities*. Oxford: Blackwell.

Chapter 10

Student Preferences for University Accommodation: An Application of the Stated Preference Approach

Harmen Oppewal, Yaniv Poria, Neil Ravenscroft and Gerda Speller

Introduction

Student accommodation has been a popular topic of research in the environment-behavior discipline (e.g. Abu-Ghazzeh, 1999; Kaya and Erkip, 2001). Commonly such studies look at a limited number of attributes of dormitories as a physical environment and relate them to certain aspects of behavior or to student preferences or perceptions. In this study, however, students' overall preferences towards student rooms are investigated by exploring various room attributes and student characteristics. More insight into student preferences is increasingly relevant in today's political climate, as universities increasingly compete to attract students, as well as finding new ways to generate revenue (Connell, 2000). One factor influencing students in their choice of university is the accommodation on offer (Connell, 2000; Murray, 1996).

The present research aims to measure students' preferences for various accommodation attributes and to determine if preferences differ between types of students. It employs a stated preference approach to obtain the measures, thus demonstrating a method that is not well known yet in the environment-behavior literature, but that has become quite popular in related fields such as transportation, housing and marketing (Green, Krieger and Wind, 2001; Louviere, Hensher and Swait, 2000). More insight into students' trade-offs between accommodation attributes could be helpful in the process of managing and designing student accommodation and in the development of optimal learning environments for students. The results of this study may also be useful for other accommodation facilities, by shedding light on occupants' preferences in relation to their personal characteristics and the potential uses of the room.

The chapter commences with a review of recent studies about the attributes of dormitories and other student-type accommodation. The methodological framework is then presented, with an outline of the steps involved in the stated preference approach and a description of the fieldwork and the site where the study took place. The results are then presented and, in the last section, some conclusions are drawn.

Literature Review

Dormitories, or halls of residence, have provided the setting for a variety of studies (Baum and Paulus, 1991). The characteristics of the population, the physical attributes of the accommodation and the long-term nature of the residence are all of interest to researchers who have investigated the relationship between the physical environment and aspects of social behavior.

A common theme in many of these previous studies was a focus on occupation density and crowding, and the effects of these factors on social behavior. Bickman et al. (1973) studied helping behavior in relation to the density of dormitories. Mullen and Felleman (1990) investigated the effects of the number of people in a room. They found out that students who had lived longer in the dormitories were less sensitive to crowding. Kaya and Erkip (2001) found that residents of higher floors perceive their room as larger, feel less crowded and hence are more satisfied than ground floor residents. Other studies concerned issues such as crowding stress (Aiello et al., 1981; Baron et al., 1976), social support (Lakey, 1989), and interpersonal relationships and personal control (Baum et al., 1979). They all suggest that students have different perceptions of the physical dormitory environment.

Studies have also looked at student preferences for particular room attributes. High and Sundstrom (1977) examined the degree to which a room's furniture could be rearranged and found that flexibility leads to a greater range of interpersonal activities than a 'non-flexible' room arrangement. Sommer (1968), in a study of four dormitories, investigated factors such as cost, privacy and study arrangements and also distance from campus. He suggested that as the distance from campus increases, students feel more isolated from the social life on campus.

Surprisingly, among all these studies, there seem to be none in which student preferences towards their room or its attributes have been investigated. Furthermore, the techniques used to evaluate student preferences have not asked participants how they compare, or trade-off, the importance of one attribute with others.

Methodology

Stated Preference Methods

Stated preference methods (also known as conjoint analysis approaches) have become increasingly popular over the last decade in disciplines such as marketing and transportation. There have, however, been only a few applications to the study of environment and behavior (e.g., Oppewal and Timmermans, 1999). Recent introductions to and reviews of stated preference methods can be found in Louviere (1988), Louviere, Hensher and Swait (2000) and Green, Krieger and Wind (2001).

Stated choice methods present respondents with experimentally designed descriptions of hypothetical objects, or choice alternatives. Respondents are asked to rate these alternatives or choose from sets of alternatives. These responses are analyzed to reveal how the different characteristics of the alternatives contribute to the overall evaluations. The approach thus involves a series of steps.

1. Identification of the relevant attributes to describe the hypothetical alternatives, including the levels over which to vary the attributes;
2. Selection of an experimental design to guide the creation of a feasible number of alternatives for respondents to evaluate. Typically fractional factorial designs are used, which allow estimation of the main effects, upon the assumption that (most) interactions can be ignored;
3. Design of the task instructions, explaining to the respondent the evaluation context or choice situation being analyzed and the response format being used;
4. Specification of a mathematical model to relate the responses to the attributes and types of alternatives;
5. Collecting respondents' responses to the alternatives, which can be a rating for each designed alternative or choices from (designed) sets of alternatives;
6. Analysis of the responses; that is, estimating the parameters in the assumed model. This is typically done through the application of some regression based approach;
7. Assessing the model performance in terms of fit and predictive ability;
8. Interpretation of the model parameters and application of the model.

The following subsections will outline how these steps were conducted for the present study.

Definition of Attributes

To find out which attributes students take into account when they choose a room, exploratory interviews were held with students of the university that was the site of this study. These students were asked which features they would recommend to new students faced with choosing a room. A variety of attributes emerged from these interviews, of which eight seemed most pertinent to include in the survey. Seven of these were included because they were mentioned most frequently. These were related to weekly rent, the number of people with whom to share toilet and shower facilities, and distance from campus. The view from the room was mentioned less often but was included because it was of direct interest to the researchers. It should be noted that one of the most commonly mentioned attributes was whether a room was shared or not. However, as the selected university offers no shared accommodation (apart from a special family and a few other rooms), it was decided not to include this attribute.

The eight attributes and their levels are listed in Table 10.1. An attempt was made to choose attributes and attribute levels that were realistic in the context of the university that was selected as the site for this study. The attribute levels were chosen based on formal documentation from the university as well as information from the interviews.

Table 10.1 The selected attributes and their different levels

- mixed or single gender floor (mixed gender, single gender)
- mixed or single course floor (undergraduates only, postgraduates only, mix of both)
- toilet and shower (private, shared by four students, shared by seven students)
- view from the room (park, building, tree)
- size of the room (9m2, 6m2, 4 m2)
- distance from campus (3 km, 1km, on campus)
- age of building (old, renovated, new)
- rent per week (£40, £50, £60)

Experimental Design, Dependent Variables and Model Specification

Because it would be impossible to present all possible (2×3^7) attribute combinations, we chose to use a fractional factorial design. Such designs allow the estimation of main effects under the assumption that interaction effects can be ignored. A design of 18 treatments was selected from the full factorial design, hence 18 different attribute profiles were created for presentation to the respondents. In the final survey instrument these 18 profiles were presented in six sets of three profiles, with two sets printed on one page. The order in which the profiles appeared in the survey instrument was randomized; this randomization was repeated three times, so there were four different versions of the questionnaire.

For each profile, respondents were asked to indicate 'how much they like or dislike' the accommodation, on a scale of 1 to 9, where 1 represents the 'worst possible room' and 9 'my ideal room'. As it was felt that this is not a common questionnaire format, an example was provided at the start of the questionnaire. An example of a profile as presented to the participants is given in Figure 10.1.

Room A
On a single gender floor
Floor shared with undergraduate students only
Toilet and shower in the room
View of a tree
6 square meters (2 × 3)
Located 3 km from campus
In a recently renovated building
£50 per week
Worst Best 1 - 2 - 3 - 4 - 5 - 6 - 7 - 8 - 9

Figure 10.1 Example attribute profile describing one possible room, with rating scale

Before receiving the experimental tasks, respondents were asked some questions about the importance of room attributes and about possible room views. After the experimental tasks, several attitudinal questions were asked, followed by some basic socio-demographic data. Some of the personal characteristics investigated were taken from literature concerning student lifestyle in general, as it was suggested that this may influence students' preferences (e.g. Abu-Ghazzeh, 1999; Bennet, 1974; Greenfield, 1997). The researchers also looked for student characteristics that accommodation management would be able to use in future allocation procedures. Several additional questions were asked in the survey but the analysis of that data is beyond the scope of the present paper.

In terms of the mathematical model, we assume that the like/dislike ratings can be described as a linear function of the manipulated attribute levels. That is, we assume that a rating V_{ij} of profile j by individual i can be explained by the following function:

$$V_{ij} = b_0 + \Sigma b_k X_{ijk} + e_{ij}$$

where X_{ijk} represent the (coded) attribute levels of attributes k (k= 1..K, K being the total number of attributes), b_0 and b_k representing parameters to estimate that represent a constant and the attribute 'weights', respectively, and e_{ij} representing a random error term.

Study Site and Sample

The present research was conducted at a university in southern England. This university has a diverse range of student accommodation and hence provides a good venue for studying student preferences. The accommodation is arranged in 'courts' around campus, with the courts varying in terms of their distance from the center of campus, the number of people sharing communal facilities on each floor (toilets, showers and kitchen), room prices and availability of cleaning services. Students can also choose to stay in mixed or single gender facilities. In some courts, students can indicate whether they wish to rent their rooms for a semester, a session or a year. The courts themselves are also designed differently: some look like flats while others resemble semi-detached houses. The age of the buildings varies: some are less than three years old while others are more than twenty years old. It is felt that this variety will enhance the possibilities to generalize the findings of this study to other universities and accommodation facilities.

After several pretests had been conducted, a total of 500 questionnaires were distributed among undergraduate and postgraduate students during the first weeks of the new semester. The present paper analyses the 152 questionnaires that had been returned within the first weeks of the semester. A majority (79%) of these participants were between 18 and 22 years old; 15% were between 23 and 26; 6% was older than 26. There were slightly more females (55%) in the sample than males. Of the sample, 78.2% were registered on an undergraduate course and 21.8% on a postgraduate course.

Most students (59%) had spent most of their life in England, 18% had spent most of their life in another European country and 15% reported spending most of their time in an Asian country. The other 7.5% spent most of their life in other places. Of the sample, 65% reported English as their first language.

Analysis and Results

Model estimation

A multiple regression analysis was conducted to estimate the parameters in the linear additive model specified above. The estimated model predicts the overall effect of each of the attributes on the participants' responses. The underlying assumption that the ratings are interval data is commonly accepted in stated preference analysis. The model fits the data reasonably well, given the disaggregated nature of the data ($R^2 = 0.21$; adjusted $R^2 = 0.20$).

Table 10.2 Parameter estimates obtained from the regression analysis

	Unstandardized Coefficients	Std. Error	Standardized Coefficients	T	Sig.
(Constant)	4.69	0.03		144.25	0.00
GENDER	-0.28	0.03	-0.15	-8.67	0.00
WITH UGS OR PGS_1	-0.21	0.05	-0.09	-4.58	0.00
WITH UGS OR PGS_2	0.16	0.05	0.07	3.49	0.00
SHOWER AND TOILET_1	-0.07	0.05	-0.03	-1.64	0.10
SHOWER AND TOILET_2	-0.59	0.05	-0.26	-12.79	0.00
ROOM VIEW_1	-0.18	0.05	-0.08	-3.87	0.00
ROOM VIEW_2	0.00	0.05	0.00	0.07	0.95
ROOM SIZE_1	0.08	0.05	0.03	1.73	0.08
ROOM SIZE_2	-0.33	0.05	-0.14	-7.09	0.00
DISTANCE_1	-0.01	0.05	-0.00	-0.18	0.86
DISTANCE_2	0.65	0.05	0.28	14.05	0.00
OLD OR NEW_1	0.17	0.05	0.07	3.71	0.00
OLD OR NEW_2	-0.03	0.05	-0.01	-0.62	0.53
RENT_1	0.01	0.05	0.01	0.29	0.77
RENT_2	-0.13	0.05	-0.06	-2.94	0.00

The results are shown in Table 10.2. The table lists the parameters as estimated for a set of indicator variables that represent the differences between levels within attributes. The indicator variables were created using effects coding, which means that for each three-level attribute two indicator variables are constructed. The first level is coded -1 for each indicator variable, the second level is coded 1 for the first indicator variable and 0 for the other; the third level is coded 0 for the first indicator variable and 1 for the second indicator variable. Hence, each indicator variable represents the difference in the means of the observed ratings for two levels of one attribute. For the two-level attribute only one indicator variable is required, which was analogously coded as -1 and 1 for the respective two levels.

Attribute Effects

All attributes have a significant effect on the dependent variable, though not all indicator variables (level differences) are significant. To facilitate the interpretation of the effects we calculated the 'part-worth' values of each attribute level from the parameter estimates in Table 10.2. Part-worth values are the relative contributions of each level to the total predicted score. The total predicted score for any profile is the sum of the part-worths of its levels, plus the constant of the regression. For example, the part-worth of level one of the first three-level attribute is -1*-0.21 + -1*0.16, which equals .37; the part-worth of the second level is 1*-0.21 + 0*0.16, or -0.21. A plot of all the part-worth values is displayed in Figure 10.2. Note that for each attribute the part-worth values sum to zero.

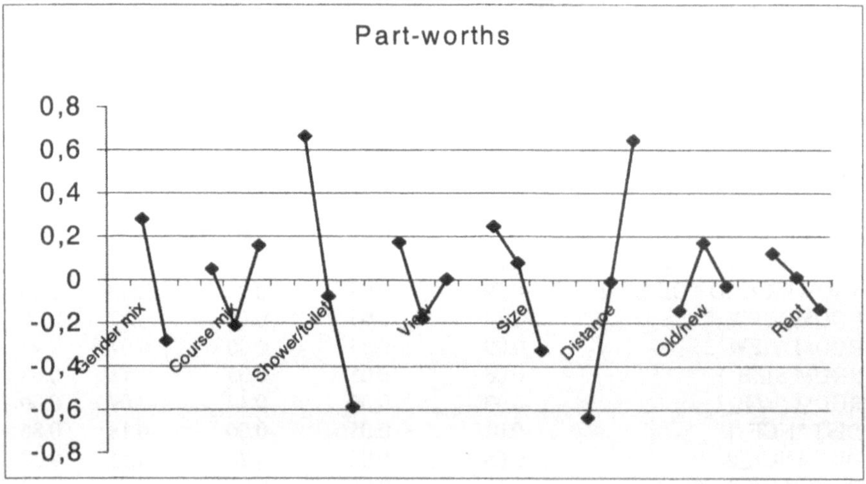

Figure 10.2 Attribute level effects as deviations from the overall mean

The figure clearly shows that the largest effects occur in relation to whether shower and toilet facilities are shared and how far the accommodation is from campus. Respondents show a strong preference for private facilities, although it should also be noted that this preference declines markedly if facilities are shared with seven instead of four students. Evidently it is not just a matter of having private facilities or not, but also the number of people with whom these need to be shared.

Respondents also show a strong preference for a room on campus, with preference decreasing with greater distance from campus. It is not just the fact of living on campus or not that is of influence, but also the distance from campus. Note that one kilometer from campus suggests a distance that is within reasonable walking distance, whereas three kilometers implies a greater dependency on means of transport like buses. (The university actually runs a free and frequent bus service for students in accommodation away from campus).

The next most important attribute is the size of the room, in terms of the difference between four and nine square meters. Within size, the largest effect is observed for the change from six to nine square meters. Then comes gender mix of the floor, with respondents showing a preference for a mixed gender floor. Further analyses reported below show this preference does not depend on the gender of the respondent.

The remaining attributes seem all of fairly equal importance. Floors with mixed occupation by undergraduate and postgraduate students are preferred over single degree level floors. Students seem to prefer renovated accommodation to accommodation in new buildings, although they do not like old buildings. The smaller number of postgraduate students in the sample can explain the low preference for floors with only postgraduate students. As shown further below, if we test the effects separately for undergraduate and postgraduate students it appears that undergraduate students in general do not like to be on a floor that has a majority of postgraduate students. Vice versa, postgraduates do not like to live on a floor with a majority of undergraduates. There is no significant difference between the part-worths of single (own) degree level and mixed degree level floors. Regarding the room view attribute, a view of a park is strongly preferred over a view of a university building, with the view of a nearby tree taking an intermediate position.

Finally, rent has a relatively small effect, considering the amounts involved. That is, the difference between £40 and £60 per week has a smaller effect than almost all other attributes. Although, in stated preference analysis, the attribute trade-offs against price (rent) are often used as an indication of willingness-to-pay, it seems that for this study there should be much caution about such an interpretation. Possible reasons for the low effect of rent are, firstly, that students may have 'gone easy' on rent because they did not consider this as their own money – for many students it is the parents who foot the bill of their accommodation costs. Another possibility is that the respondents did not fully take price into account when giving their 'liking' ratings for the profiles, their ratings expressing the inherent quality or value of the room, not value for money. Finally, it is possible that students indeed were willing to pay substantially more for improvements to their accommodation. Although some of our pilot interviews seem to support this possible explanation we are hesitant to draw such a conclusion without further evidence.

Student Differences

We next tested if the preference differed with classification variables that are important for the campus accommodation management. Effects were tested separately for gender, degree level and country of origin, by including the interactions of the variable with all the attributes in the regression model. The model showed no improvement in fit when the interactions with gender were added ($F=1.257$, $df=15, 2590$, n.s.), although there was a significant improvement with degree level ($F= 7.125$, $df= 15, 2590$, $p<.001$). Because all added effects are orthogonal to the effects already included in the model, we only display the extra parameters (see Table 10.3). Significant effects show up for gender mix, degree level mix, showers and toilets, and old or new building. It appears that undergraduate students are much more interested in being on a mixed-gender floor than are postgraduate students. Undergraduate students are also keen to not be on a floor with only postgraduate students, and prefer a renovated to a new building. Postgraduates find it much more important than undergraduates to have their own shower and toilet, or to share these facilities with fewer people. There are no significant differences in terms of preference for room views, sizes, distance from campus or weekly rent.

Table 10.3 Parameter estimates for interaction effects of attributes with respondent's degree level

INTERACTIONS OF DEGREE BY:	Unstandardized Coefficients	Std. Error	Standardized Coefficients	t	Sig.
GENDER	-0.19	0.04	-0.11	-5.08	0.00
WITH UGS OR PGS_1	-0.25	0.05	-0.11	-4.59	0.00
WITH UGS OR PGS_2	0.08	0.05	0.04	1.53	0.13
SHOWER AND TOILET_1	0.12	0.05	0.05	2.15	0.03
SHOWER AND TOILET_2	0.24	0.05	0.10	4.38	0.00
ROOM VIEW_1	-0.05	0.05	-0.02	-0.95	0.34
ROOM VIEW_2	-0.01	0.05	-0.00	-0.23	0.82
ROOM SIZE_1	0.08	0.05	0.03	1.42	0.16
ROOM SIZE_2	-0.05	0.05	-0.02	-1.01	0.31
DISTANCE_1	0.02	0.05	0.01	0.47	0.64
DISTANCE_2	-0.01	0.05	0.00	-0.00	1.00
OLD OR NEW_1	0.13	0.05	0.06	2.39	0.02
OLD OR NEW_2	-0.06	0.05	-0.03	-1.19	0.23
RENT_1	-0.11	0.05	-0.05	-1.92	0.05
RENT_2	0.02	0.05	0.01	0.41	0.68

Finally we tested effects for country-of-origin of the student. This variable highly correlated with the degree level as a majority of the postgraduate students at this university are from overseas. We therefore tested the country-of-origin effect within the group of undergraduate students and found only a significant effect at the alpha is ten percent level ($F=1.582$, $df=15, 2053$, $p<.07$) which can be totally attributed to the effect of mixed gender floors. Among the undergraduate students, students from the UK have a much stronger preference for mixed gender floors than students from elsewhere (estimated interaction parameter of country of origin and gender mix is -.114; $p<.001$).

Conclusion and Discussion

This research aimed to measure how room attributes influence student preferences towards university accommodation. Results obtained from a sample of undergraduate and postgraduate students at one university in the UK show that the students are most sensitive to whether they need to share shower and toilet facilities with other students and how far their accommodation is from campus. Room size (four or nine square meters) is the next most influential attribute, followed by the mix of gender and mix of undergraduate versus postgraduate degree level students on the floor of residence. The view from the room has a smaller, although still significant effect. Weekly rent had a surprisingly small effect in this study. This may indicate a willingness among students to pay substantial amounts for improvements to their rooms. We are however hesitant to draw this conclusion without further evidence. Possible alternative explanations for this effect are that respondents ignored the value-cost trade-off when expressing their 'liking' for the accommodation profiles or that rent is not their primary concern, as often parents pay this cost. Regarding student differences, it was found that, compared to postgraduates, undergraduate students show a strong preference for mixed gender floors and renovated instead of new buildings.

The results presented here can be used to support the management student accommodation in terms of creating a better fit between students' accommodation preferences and the type of accommodation offered to them. This can result in a more efficient use of accommodation as well as higher levels of student satisfaction. The results can also inform the design and planning of new student accommodation, for example in terms of how accommodation at a greater distance from campus can be made sufficiently attractive – for example by providing more private facilities and larger rooms. Providing a more pleasant view from the room might also help.

The results are based on a data collection at only one university. Future research in other settings would provide more information with respect to the possibility of generalizing the findings of this study. Also, the scenarios presented to the students were built around eight attributes, each having two or three levels. Not all attributes that students take into account were investigated, and not all possible levels were included. Moreover, the experimental design used here did not allow measuring the effects of particular combinations of attributes or attribute levels. It may be that some attributes only have an effect in combination with other attributes. These are clearly topics for further research. Despite these limitations we hope this study has demonstrated the usefulness of stated preference methods for environment-behavior research.

Acknowledgement

This research was part of the EU Framework V funded 'Greenspace' project (www.green-space.org).

References

Abu-Ghazzeh, T.M. (1999). Communicating behavioral research to campus design: factors affecting the perception and use of outdoor spaces at the University of Jordan. *Environment and Behavior*. 31(6), 764-804.

Aiello, J.R., Baum, A., and Gormley, F.P. (1981). Social determinants of residential crowding stress. *Personality and Social Psychology Bulletin*. 7(4), 643-649.

Baron, R.M., Mandel, D.R., Adams, C.A., and Griffen, L.M. (1976). Effects of social density in university residential environments. *Journal of Personality and Social Psychology*. 34(3), 434-446.

Baum, A., and Paulus, P.B. (1991). Crowding. In D. Stokols and I. Altman (eds.). *Handbook of Environmental Psychology*. 533-570. Malabar: Krieger Publishing Company.

Baum, A., Shapiro, A., Murray, D., and Wideman, M.V. (1979). Interpersonal mediation of perceived crowding and control in residential dyads and triads. *Journal of Applied Social Psychology*. 9(6), 491-507.

Bennett, D.C. (1974). Interracial ratios and proximity in dormitories: Attitudes of university students. *Environment and Behavior*, 2(2), 212-232.

Bickman, L., Teger, A., Gabriele, T., McLaughlin, C., Berger, M., and Sunday, E. (1973) Dormitory density and helping behavior. *Environment and Behavior*. 5(4), 465-489.

Connell, J. (2000). The role of tourism in the socially responsible university. *Current Issues in Tourism*. 3(1), 1-19.

Green, P.E., Krieger, A.M. and Wind, Y. (2001). Thirty years of conjoint analysis: reflections and prospects, *Interfaces*. 31 (3, Part 2 of 2), S56-S73.

Greenfield, T.A. (1997). Gender-and grade-level differences in science interest and participation. *Science and Education*. 81(3), 259-276.

High, T., and Sundstrom, E. (1977). Room flexibility and space use in a dormitory. *Environment and Behavior*. 9(1), 81-90.

Kaya, N., and Erkip, F. (2001). Satisfaction in a dormitory building. The effects of floor height on the perception of room size and crowding. *Environment and Behavior*. 33(1), 35-53.

Lakey, B. (1989). Personal and environmental antecedents of perceived social support developed at college. *American Journal of Community Psychology*. 17(4), 503-519.

Louviere (1988). *Analyzing Decision Making*. Sage University Paper Series No. 67. Newbury Park, CA: Sage.

Louviere, J.J., Hensher, D.A., and Swait, J.D. (2000). *Stated Choice Methods: Analysis and Application*. Cambridge: Cambridge University Press.

Mullen, B., and Felleman, V. (1990). Tripling in the dorms: A meta-analytic integration. *Basic and Applied Social Psychology*. 11(1), 33-43.

Murray, M. (1996) Campus accommodation. Hospitality, 153, 32-33.

Oppewal, Harmen and Timmermans, Harry J.P. (1999). Modeling Consumer Perception of Public Space in Shopping Centers. *Environment and Behavior*. 31 (1), 45-65.

Sommer, R. (1968). Student reactions to four types of residence halls. *The Journal of College Personnel*. 4, 232-237.

Chapter 11

Culture and Architecture: Theoretical and Methodological Issues

William J. Thompson

Introduction

Imagining the universe as a 'material and mechanical world' has become one way of articulating an ecological model of the world. We have become used to the idea of knowing as a way of thinking about a model in particular ways, often unique to one particular kind of knower (Fuller, 2000). It allows us to confirm certain parts of 'our' world and to deal with parts that are the subject of controversy by agreeing to know in different ways. However problems arise when exploring segregated ways of knowing, crossing boundaries, such as for example when considering people-environment matters. There are significant difficulties when referring, for example, to 'embodied' as the body configuration of what is known. This is because the idea of the knower and the known includes the concept of a linkage between the body and surroundings. The idea of knowledge is perhaps the link itself, a phenomenon of experience rather than either the nervous system or the 'object in space' that is cognized. Thus the body and surroundings are materially part of a teleology that exists as a category or set of contextual relationships communicated amongst a membership of individuals.

The teleology demands specific behaviors towards contextual relationships that are not required of non-members and may be invisible to them. The solution adopted by many in the field of environmental design is to build bridges between disciplines, to network, form groups, develop translations. This 'utilitarian' enthusiasm for bridge building and networking is helpful to any who desire to communicate as full members of each others' disciplines. Those who cross over seek to resolve controversy previously 'allowed for' as difference by sharing sophisticated concepts and rules that allow transactions to occur between them as well as between the original membership groups. Thus there is a kind of trade between disciplines that allows an illusion of group exploration as if it is cross boundary possibly leading to the formation of a new segregation: a kind of group relativism adding to the managerial layer of production. In effect a new group, a new boundary, has been formed with new, improved, contents[1].

[1] Coyne R. (1996) refers to utilitarian instrumentalist and liberal sceptics in relation to professional education.

The Architect's Problems

In 1969 the psychologist David Canter (Canter, 1969) suggested that architects needed to know more about psychology[2]. Initially in the UK the term for a 'new approach to' the way buildings affect people was 'Architectural Psychology' (Canter, 1969, p.11) whilst in the USA it was Environmental Psychology (Ittelson, 1976). Architects claim to be trained to embody the skills required to configure buildings and hence make claims to control the shape of any built environment. Spatial knowledge and skills in the workplace can acquire epicenters of cost, time and/or quality of the building or some concept of architectural service (Duffy, 1998). However critics of the architectural claim deny that architects have any kind of acceptable theory of the built environment in a way that links sentient people together as part of a life-world experience (Lefebvre, 1994; Anderson, 1998).

Architects do sometimes encourage questions about the use of buildings by people (Zimring, 1987; Kernohan, 1992); and about the design of buildings for people (Zeisel, 1984). Those making claims that they do, after all, pay attention to the quality of life provided by the building imply, but do not describe, the inclusion of some kind of link between one and/or a number of mentalities and physicality. The theoretical position of J. J. Gibson (Gibson, 1974) is often called upon to justify claims linking thought to things. Gibson hypothesizes links between surroundings and 'mental contents' that could be thought of as being shared: this has a century long tradition in psychology (Heft, 2001). Such claims as are made demand some way of understanding how we might put buildings into the heads as much as we seem to have mastered our ability to put heads into the buildings (Mace, 1977).

Discussing Theory

The questions we should be asking are, for example, how do 'thought-thing' links come to be? Is it likely or inevitable that surroundings will give rise to phenomena that stimulate nervous activity in such a way that it is exactly similar in all humans having the same sense organs? Structuralist theories flourished during the 20th Century, promoting the application of systems thinking to behavior (particularly suited to the rationalist concept of brain as machine) but these are still very much part of theories (Lawrence, 1993) that ignore evolutionary concepts. Jane Jacobs (Jacobs, 1963) held that environmentalists failed to deal adequately with the problems of consciousness. In effect we are failing to deal with other unique 'thinking' people in space. These 'others' are just as likely to have very different as have very similar experiences in their specific life-world.

The philosopher Chalmers (1997) describes three categories of theory - A, B and C, relative to sentient beings in a material world. Theory A denies consciousness absolutely; Theory B extends consciousness into the individual.

---

[2] The place of architectural psychology in proceedings of the architectural psychology conference at Kingston Polytechnic in September 1970 p.3.

Both are mechanistic in the way that Harvey and Descartes conceived the human to be a machine 'designed' for some purpose or other and interacting with the world on a cause and effect basis. Chalmers calls theories A and B 'easy options' because they describe functionality and deny the hard part, which for him is describing mental space as being occupied along with physical space. We live in the physical space and are conscious of it in the mental space.

Many environmental disciplines believe that buildings or relationships of one kind and another will do things to people, that there are rules and laws linking people and buildings, that they can design for a purpose. In believing that they remain resolutely within the boundaries of theories A and B and refuse to move outside of these boundaries.

Memetics

Dawkins coined the term memetics to bring biology and pure science (people and things, thought and things) closer together in theory (Dawkins, 1989; Aunger, 2000). His original idea was to explain a mechanics of valuation by the senses. Our perceptual mechanisms are able to value what surrounds us without us being consciously aware of making that valuation. Input can be brought to our conscious attention by a combination of mechanisms that are generally heritable and modular (Churchland, 1995). This relates comfortably with Edelman's concept of epicentered consciousness (Edelman, 1992).

Hypothetically, the perception of the nervous system produces the meme. It seems that the individual can engage with surroundings as if each is tailor made for the other - such is the flow of interaction between occupants and space (Seamon, 1979). On linking perception and object we conjecture that a ballet of movement rest and encounter is something every individual may experience and even control just as we know they are capable of holding conversations and writing meaningful statements.

Theories A and B (above) require objects or symbols (some sort of boundedness to experience) in rather the same way that sign and symbol are understood in semiology (De Saussure, 1983; Peirce, 1960). Several architectural approaches to spatial cognition use these theoretical platforms (Hillier, B. and Hanson, J., 1990; Lawson, 2001; Scruton, 1994). In one of the more cognitive papers, treating experience in space, space syntax methodology has used the concept of bounded spaces and Gibson's work on visual perception (Peponis, 1997). The point of space syntax methodology, however, is to mathematize the spatial relationships so as to retrieve information more easily, a sort of GIS approach. Thompson has criticized existing space syntax methods for their exclusion of 'theory C' (Thompson, 2001) and is dubious about Gibson's proposal in full (Thompson, 2003).

The problem is that any system, used to represent any life-world experience, has a propensity to treat the system it utilizes as if the individual pays conscious attention to it or that the boundary 'used' to contextualize parts of the system fixes itself. This criticism was first understood simply as a problem of developmental exposure to the spaces (Thompson et al., 1998) but shifted to a problem of perception and action (Thompson, 1999).

More explicitly the problem is that by relying upon geographical or mental boundedness and a naïve version of the nervous system linking surroundings to mental manipulation (Hart and Moore, 1973) we create an epiphenomenon for each phenomenon in the life-world. In its crudest sense J. J. Gibson's affordances permit this interpretation. It is a regression that seems plausible as an explanation of what is a link between mental contents and surroundings – affordances make use of this – but it is not an explanation of how the person-environment link comes in and out of being, just an acknowledgement of it.

Clearly the human species demonstrates a propensity for language (Bickerton, 1996) and we may justifiably claim that we build languages and operate grammars as part of the human life-world (Uexkull, 1957). The nervous system does exhibit cause-effect relationships with surroundings (Sherrington, 1973) and a propensity to use certain spatial features within those relationships (Lindberg, 1984) but the feature does not dictate its use without the user making choices (Carr and Schissler, 1969). Movement, for example, clearly has evolutionary developments selected for biologically (Fox, 1971) and habit or practice alters behavior (Berthoz, 2000). However some individuals pay attention to things that other people ignore and by paying attention can develop distinct skills that emerge as differences between them such as explorer 'types' and non explorer types (Hazen, 1982). So can we choose our affordances? It certainly seems that we all have abilities close to that.

The point here is that although we can reduce Euclidean space to make it easier to count (quantify) that does not mean that individuals actually count spaces in any reliable fashion. We can, it seems, use the perceptual mechanism and nervous system to remove ambiguity and agree unequivocally and thus assist discourse. But the fact of language it is not an explanation of what a relationship between the person-environment does and it is the shift from the specific to the general that is uncomfortable when psychologists talk to, say, architects who are perhaps looking for universal 'meanings'. As suggested by Wapner (1980), it is neither space nor humanity but phenomenological experience which eventually decides what is available. In other words 'links' are reductions of possible meaning that conceal a more profound and extensive 'reality' that is felt phenomenologically – any communication is bound to be reductive because it reduces both space and time in the 'act of' communication.

Tolman's Rats

Experience includes what seems to be non-conscious selection of what we can call elements (Tolman et al., 1946) that can be imagined as a combination of material and neural configurations, allowing certain aspects of environmental behavior relative to those elements to come into being (LeDoux, 1996). These kinds of behaviors arrive in such a way that a person becomes conscious of that behavior rather than selecting for it consciously, hence the surprise we feel when we become conscious of, for example, foveating (staring) at someone in an embarrassing way.

This suggests that consciousness can draw upon experience as water from a well of experience filled by both considered and serendipitous action across a range of parallel or simultaneous or symbiotic inputs. Selecting for and sustaining an invariance in behavior as a conscious part of experience becomes a way of sustaining behavior as an invariant condition of communicating by doing (Barker, 1973) and can be acquired by a process of trial and error over time. There is a notion of immersion within experience as well as that of an imagination that can bring into being alternatives to what is perceived and or expected by others.

Tolman's rats seemingly noticed more about their own life-world (the position of lit lamps in the room, for example, and relationships between lamp and food) than simply the route 'forced upon them' during an experimental run. The configuration of the rats' central nervous systems was not only related to the lab technician's view of the world. Each rat's record of experience came from the richness of the relationship between the fact that the lamp existed and for example the cognitive propensity of each individual rat. What each 'chose' to pay attention to (rat decisions were variable) - the 'lamp' and the food - as 'historic' elements placed into a new construct, the short cut, for example; the rat 'doing its thing' to get at the food.

Tolman thought this interesting because it implies that each rat creates its own synchrony, located in the same space and time as humans yet different. Selection was made from a complex integration of elements capable of being isomorphic[3] in some way yet seemed to respond to contextual changes rather than being an entirely mechanical link such as that which Gibson's affordances lend themselves to. Tolman's work was sidetracked[4] into the concept of mental or 'cognitive' maps (Kitchen R, 2000) in the search to link thought to things by way of mental maps

[3] Having a one to one relationship between the two phenomena, the nervous system and surroundings so that what changed in one also 'caused' change in the other in an exactly proportionate but not necessarily identical manner. I.e.; the two systems may be differently constructed but each slavishly follows the others actions - in this sense of the term at any rate.

[4] By saying sidetracked I mean to imply that Tolman's work got quite passionate about personal freedom and the right to choose and to live within a rich environment using a rich palette of experiences (the liberal attitude perhaps); however, the use of cognitive mapping does not move down that road, preferring to identify and represent only those isomorphic linkages between body and surroundings or to ask questions about the lack of isomorphism - Tolman accepted this transformational aspect of isomorphism as a 'natural' product of the 'nervous system in the human life-world'.

that might correlate with geographical maps. This works as a language system, provides the occasion for discourse, encourages group belief, but fails to answer Chomsky's question of where language came from? (Chomsky, 1957) (assuming at that time it was innate, later revised somewhat (Aitchison, 1996)).

Just two decades earlier Tolman had asked a similar question about environmental behavior. Tolman suggested that we might refer to *rich* or *impoverished* environments. He used the terminology of 'narrow' and 'wide' band experience and created the analogous use of maps as representative of the memory of an environment in the wider sense, opposing this with narrow or impoverished memory as that of only the necessary route or way through any configuration to a goal (Tolman, 1938).

Implications for Architectural Order

The principle feature of cognitive theories A and B (above) is their use of the mechanical model, a reductive model in its mechanical form, of the way people relate to space and time (movement through space) so that believers in theories A and B can get on with being experts at what they do.

Such models include the mechanics of culture as species, i.e. include the people in space as 'having' cultural rules and 'being' cultural (Hall, 1977), but do not let us understand why or how the culture arrived in the first place (Tooby and Cosmides, 1992). Even the use of the term model implies a mechanics which relies upon an isomorphic perception action cycle (Brodbeck, 1969) and this then, in turn, upon the mechanical biology of Harvey and Descartes (Lewontin, 2000).

Cognitive theories A and B have to maintain a specific relationship between some kind of object related perception and some kind of modular processing by the central nervous system resulting in a perfectly formed and mechanical link between outside and inside - which is simply not the case nor even a close model of what is the case (Fodor, 2000).

The notion of immersion, on the other hand, in some fluid relationship that phenomenology can include quite easily (Csikszentmihalyi, 1988) reflects the same approach as that of the gene pool in genetics in terms of how change may occur whilst also allowing a mechanical process its development. The variant between the momentary and long-term provide us with the illusion of a mechanics 'acting on' our life and those of other people (Bowlby, 1969). This version of selection and experience is harder to contemplate than the usual concept of spatial cognition used in architectural schools, basing their activities on theories A and B (above) that require individuals, for example using their proposed buildings, to act 'normal' or allow themselves to be considered odd. We can relate physical space to mental space as having attributes contingent upon its delivery by perceptual input – the so-called bottom up side of perception and the opposite, top down side of perception provided by the integrating action of the life-world (Canter, 1977, p.1) but this puts the individual into a cognitive position of 'knowing' rather than one of 'coming to know' which is subtly different.

Architectural People-Space

A society which values certain kinds of behavior over others, working for money rather than stealing it, for example, has to theoretically understand various kinds of relationships between individuals as normative or default conditions for that society. Individuals are not necessarily or even desirably (for some) conscious of relationships between all individuals (Wirth, 1964). The ability to manufacture (value) systems is all things being equal a general fact of the human life-world, however the ability to share and/or tolerate value systems is probably an advanced skill that one acquires with experience that is wide enough to provide an individual with conflicting values for the same set of elements. Such systems are complex, difficult and 'fragile' at some stages of development and robust, simple and easy at others (Giddens, 1991 p.24).

Rather than a mechanical nervous system or rational thinking machine (also known as a computational model) there is a credible alternative view of the individual as a relativity machine (Harre, 1998; Young, 1994). This is the ability to feel the effect of utilizing a knowledge set with others and of creating such a set for personal and/or altruistic 'reasons'. We have evidence to show that the environment helps to configure the nervous system (Blakemore, 1998) and emotional relationships to surroundings alter neural connectivity and capacity for selective action (Llinas, 2001) by way of both individual and collective mechanisms. Theories which appreciate the interactivity between large structural phenomena, building blocks and cognitive agents (Holland, 1998) appreciate the possibility of developmental constructions evolving symbiotically as bottom up and top down in a social model where top down is provided by the rigid rules and bottom up is actually a relaxation of the rule rather than the intake of perception since that is also part of the top down perception action process!

Mechanisms which force the body to pay attention to change, for example to foveate, to focus the central retina upon an item, are genetic mechanics first and specification of what is chosen second (Lewkowiez, 1994). However the individual may be conditioned at an early stage in its 'life' to *position itself* more to its integrated and conditioned store of experience than the perceptual input arriving from its location in space. Tolman's idea of a rich or impoverished life-world - a relationship exhibiting 'diverse' or least diverse possibilities might be hedonistic as well as potentially socially controlled or at least controlled by way of the communicative possibilities rather than the surroundings as things. It becomes a balance between certainty and infinite possibility rather than a choice between a set of known things.

The so called top down perceptual mechanisms come to act on the input not necessarily consciously but as part of the integration of the nervous system (Dawkins, 1998; Gazzaniga et al, 1979). This bottom up-top down integration may create a dilemma which the nervous system then has to resolve. The resolution of dilemma appears to depend upon the whether or not the experience of the individual considers the situation to be a dilemma rather than any absolute rationality (Damasio, 1995). However the mechanics is heritable and 'solves' dilemmas resulting from the agent's life world experience in the human life-world as well as the larger ecological impact and geometrical perception.

Transparency and Flow

In this hypothesis consciousness emerges out of an infinite possibility for self and other that can be reduced to a communication yet still retain possibilities unimagined (Levinas, 1996). In Order to synchronize behavior with 'each other' however there must be a psychological mechanism to do so (Plotkin, 1994) and an invariant Euclidean relationship in both space and time, something for the 'psychology' to exert an action upon. The suggestion by Boulding in the preface to Zeleny's work on intellectual management (Zeleny, 1980) is that features of dynamic interaction are such that large scale planning appears to have taken place - whilst in fact no plan has been made. The relationship between individual and surrounding in Boulding's model is as if a link value relationship emerges out of interactions that lead to that inference rather than actually plan to do so. The macro environment is unplanned - selections made are contingent upon an unplanned yet ostensibly material relationship (Edwards and Tversky, 1967).

Reid (Reid, 1998) talks about 'reward cycles' relating to behaviors that appear to demand insight or some kind of overview and yet the cycles are shown to emerge without any macro or global plan. The individual has a quite mechanical relationship between itself and surrounding phenomena, in other words it is rational in its actions, but quite individual. Thus individuals can appear to be part of a grand plan, even think of themselves as part of some huge experiment, such as they do in 'movements' or 'sects' but they could be 'blind' to the larger pattern and still produce the same effect.

A point raised in Sartre's (Sartre, 1981) work is that the position of the 'conscious self' is very ambivalent. That new knowledge known to the newly conscious is a becoming both of knowledge itself and its incorporation into the knower (Dewey, 1949). Against Sartre's model, it seems as if knowledge should be viewed as the problem of consciousness being an individual experience in a state of flux yet set within the human life-world at large that has to communicate using material form, words, signs and entities of some substance or other. Godwin's model of anarchy as a unique personal experience contrasted to rules and customs without that same investment of individual living (Woodcock, 1946).

Perhaps the neatest way to appreciate how the nervous system and any unconscious extension of it into the surroundings might handle unconscious processing, whilst leaving it open to change, is the analogy of the computer software and use of the default mode of operation. We can consider our own nervous system as symbiotically forming a default representation (Reason, 1992) if only from the impetus of novelty incoming, thus we speed up the process of moving in the environment if we build a representation mostly from memory (80%) rather than from our surroundings (Dawkins, 1998) - it makes sense to leave ourselves free to choose as little as possible otherwise we would never get anywhere or do anything in a hurry.

A non-usual sensation is likely to trigger the time to take a trial and error check, search and find action, cognitive search (hypothesis) (Tolman, 1938) whilst the need to remove ambiguity in order to move at all drives us toward a new (and possibly altered) default. This fits very well with what neuroscience tells us about our nervous system - it codes for self, social and environmental relationships (Frackowiak, 1997) that are as discussed above reductive of the actual self, the actual social possibilities and the actual environment in its infinite scale (but not necessarily in its immediate surroundings scale). The nervous system provides its carrier with a unique personal valuation of surroundings (Churchland, 1995) but it only needs to do so enough to cope with the immediate in both space and time. We need to consider contingency in order to achieve values in common on a sustainable basis. There will be an interactive cycle of vicarious trial and error, searching for cause, hypothesizing and arrival at a suitable default by agreement and communication. The effect is to reduce or remove choice by making selections we try to maintain over time.

In addition to being 'provided with default conditions' we quite obviously have the capacity to create diverse kinds of default condition for 'ourselves' by combining surroundings and actions in different ways but we then need to articulate these if they are to become known by others in the same way. The, again obvious, possibility is that each and every individual is different from every other individual and is either forced, persuaded or chooses to understand in a similar way, at least in limiting the range of experiences which 'matter'.

The Mechanics of Natural Selection

The fixing of the configured mental relationships seems necessary in order to have recognition and sustainable development or speed of action. Altering exposure and presentation offers almost unlimited opportunities for mutation however the effort to activate or inhibit such changes has to overcome the default mechanism of other individuals and the amount of work needed to bring about change. The attention people pay to their surroundings can be seriously affected by what they paid attention to last time (Lustig and Hasher, 2002) and what they think they are searching for. The construction of the self (Cushman, 1990) might be construed as the choice one has to belong to a space and time relationship with others who share and/or exhibit certain values or meanings (and a possibility of avoiding others who do not) and thus not one self but many types of participant, learner, ignoramus and expert.

Default systems, large defaults, link the totalized and infinite in a historic combination of actions and surroundings shared by other knowers who may at almost any time alter their status within the shared relationship. The relationship between newcomers would have to be controlled as indeed we see that it is in a large number of social activities. People are shown 'the way to do things' by experts and strongly 'encouraged to conform' unless there are 'reasons' to explore relationships such that individuals invent and circumstances prevent and so on.

Summation

There can be an apparently slow transformation of the whole brought about by selection at the interface between certain default communicative groups which creates an ecological default, or the sensation of that, within the individual as if in the whole system. That transformation does not apply equally to all locations within the system because it depends upon the individuals as individuals and the communicative groups and how they interact with other synchronous activities. In fact the system is such that it ceases to be one system and becomes several synchronizations with varying degrees of tolerance and fit. Eventually the number of systems may be so large that we can stop thinking about systems and start thinking about individuals with the capacity to uncouple themselves from place and group in order to re-couple elsewhere. These individuals assemble in order to form systems and thus transformations occur which may impact upon the whole in extra-ordinary ways or may pass unnoticed because a larger and single system is illusory as much as it is real in the imagination.

This chapter has conjectured how an individual can be considered as part of the flux in the phenomenon known as experience and how 'cultures' may interact with individuals only by way of the behavior communicated or invented by groups of individuals in the short and long term. Evolutionary theory suggests that both these things are happening at every moment because the selection mechanism is neither one single universal system nor individual yet affects both as a social way of knowing.

That paradox, the evolutionary mechanism, is suggested as the basis of any person environment study. The role of the individual (architect) becomes understanding the paradox and articulating it. It also implies that phenomenology is at the heart of understanding because it is only out of that flux that alternative or ready made material elements can emerge and be acted upon in space and time. As Ken Frampton wrote (Frampton, 1996, p.9) half of architecture is about building, the other half is about consciousness. It is suggested that consciousness is not of things about which choices can be made but about immersing oneself into experience and articulating the differences between the reproduction of behavior for an audience (present or imagined) and the production of it. In this the product is teleological and acted upon by the speed and 'certainty' of the reproductive as against the uncertainty and delay of novelty and our ability to manage our imaginations around those opportunities.

References

Aitchison J. (1996). *The Articulate Mammal. London*: Routledge.

Anderson P. (1998). *The Origins of Postmerdernity.* New York: Verso.

Aunger, R. (ed) (2000). *Darwinizing Culture.* Oxford: Oxford University Press.

Barker, R. G. and Schoggen P. (1973). *Qualities of Community Life.* San Francisco: Jossey Bass.

Berthoz, A. (2000). *The Brain's Sense of Movement.* London: Harvard University Press.

Bickerton, D. (1996). Language and Human Behaviour. London: University College Press.
Blakemore, C. (1998). How the Environment Helps Build the Brain. In B. Cartledge (ed), *Mind Brain and Environment*, (pp.28-55). Oxford: Oxford University Press.
Bowlby, J. (1969). *Attachment and Loss.* London: Tavistock Institute of Human Relations.
Brodbeck, M.(1968). *Readings in the Philosophy of the Social Sciences.* New York: Macmillan.
Canter, D. (ed) (1969). Architectural Psychology Proceedings of conference held at Dalandhui. University of Strathclyde.
Canter, D. (1977). *The Psychology of Place.* London: Architectural Press.
Carr, S. and Schissler, D. (1969). The City as a Trip. *Environment and Behaviour*, 1(1), 7-36.
Chalmers, D.J. (1997). Moving Forward on the Problem of Consciousness, *Journal of Consciousness Studies*, 4 (1), 3-46
Chomsky, N. (1957). *Syntactic Structures.* The Hage: Mouton.
Churchland, P. M. (1995). *The Engine of Reason The Seat of the Soul.* London: MIT Press.
Coyne, R. (1996). Deconstructing the Curriculum. *Edinburgh Architectural Research.* Edinburgh University, 23, 144-173.
Csikszentmihalyi, M. (1988). *Optimal Experience.* Cambridge: Cambridge University Press.
Cushman, P. (1990). Why the Self if Empty. *American Psychologist*, 45(5), 599-611
Damasio, A. R. (1995). *Descartes Error.* London: Picador.
Dawkins, R. (1989). *The Selfish Gene.* Oxford: Oxford University Press.
Dawkins, R. (1998). *Unweaving the Rainbow.* London: Allen Lane.
De Saussure, F. (1983). *Course in General Linguistics.* London: Duckworth.
Dewey, J. and Bentley, A. F. (1949). *Knowing and the Known.* Boston: Beacon Press.
Duffy, F. (1998). *Architectural Knowledge.* London: E FN Spon.
Edelman, G. (1992). *Bright Air Brilliant Fire.* London: Allen Lane.
Edwards, W. and Tversky, A.(1967). *Decision Making.* Harmondsworth: Penguin Books.
Fodor, J. (2000). *The Mind Doesn't Work That Way.* London: MIT Press.
Fox, R. et al. (1971). *The Imperial Animal.* London: Secker and Warburg.
Frampton, K. (1996). *Modern Architecture.* London: Thames and Hudson.
Fuller, S. (2000). *Thomas Kuhn.* London: University of Chicago Press.
Frackowiak, R. (1997). *Human Brain Function.* San Diego: Academic Press.
Gazzaniga, M. S. and Le Doux, J. E. (1979). *The Integrated Mind.* New York: Plenum Press.
Gibson, J. J. (1974). *Perception of the Visual World.* London: Greenwood Press.
Giddens, A. (1991). *Modernity and Self Identity.* Cambridge: Polity Press.
Hall, E. T. (1977). *Beyond culture.* Garden City, NY: Anchor Press/Doubleday.
Harre R. (1998). *The Singular Self.* London: Sage Publications Inc.
Hart, R. A. and Moore, G. T. (1973). The Development of Spatial Cognition. In R.M. Downs and D. Stea (eds) *Image and Environment.* Chicago: Aldine.
Hazen, N. L. (1982). Spatial Exploration and Spatial Knowledge. *Child Development* 53, 826-833.
Heft, H. (2001). *Ecological Psychology in Context.* London: Lawrence Erlbaum.
Hillier, B. and Hanson, J. (1990). *The Social Logic of Space.* Cambridge: Cambridge University Press.
Holland, J. H. (1998). *Emergence From Chaos to Order.* Oxford: Oxford University Press.
Ittelson, W. H. (1976). Some Issues Facing a Theory of Environment and Behaviour. In H.M. Proshanksky et.al. (eds.) *Environmental Psychology* (pp. 51-59). Austin: Holt, Rinehart and Winston.
Jacobs, J. (1963). *Death and Life of Great American Cities.* New York: Random House.

Kernohan, D.(1992). *User Participation In Building* Design and Management. Oxford: Butterworth Heinemann.
Kitchen, R. and Freundschuh, S. (eds) (2000). *Cognitive Mapping*. New York: Routledge.
Lawrence, R. J. (1993). Reinterpretation of Cognitive. Institutional and Material Structure in an Integrative Historical Perspective. *Quarterly Newsletter of the Laboratory of Comparative Human Cognition*, 15(1), 16-23.
Lawson, B. (2001). *The Language of Space*. Oxford: Architectural Press.
LeDoux, J. (1996). *The Emotional Brain*. New York: Simon and Schuster.
Lefebvre, H. (1994). *The Production of Space*. Oxford: Blackwell.
Levinas, E. (1996). *Totality and Infinity*. Pittsburgh: Duquesne University Press.
Lewkowiez, D.J. and Lickliter, R. (1994). *The Development of Intersensory Perception*. New Jersey: Lawrence Erlbaum.
Lewontin, R. (2000). *It Ain't Necessarily So: The Dream of the Human Genome and Other Illusions*. London: Granta.
Lindberg, E. (1984). Acquisition of cognitive maps of large scale environments. Sweden: Doctoral dissertation, department of psychology, University of Umea.
Llinas, R. R. (2001). *I of the Vortex*. Massachusetts: MIT Press.
Lustig, C. and Hasher, L. (2002). Working Memory Span. The Effect of Prior Learning. *The American Journal of Psychology*, l.115(1), 89-102.
Mace, W. M. (1977). James J. Gibson's Strategy for Perceiving. In R. Shaw and J. Bransford (eds.). *Perceiving, Acting, and Knowing*, pp. 43-65. New York: Lawrence Erlbaum.
Peirce, C. S. (1960). Elements of Logic. *Collected Papers Vol II*. Massachusetts: Belnap Press.
Peponis, J., Wineman, J., Rashid, M., Kim, S. and Bafna Sonit. On The Description of Shape and Spatial Configuration Inside Building. Proceedings of the First International Symposium of Space Syntax, London, 1997, Vol III 40.01-21.
Plotkin, H. (1994). *The Nature of Knowledge*. London: Allen Lane.
Reason, J. (1992). *Human Error*. Cambridge: Cambridge University Press.
Reid, A. K. and Staddon, J. E. R. (1998). A Dynamic Route Finder for the Cognitive Map. *Psychological Review*, 105(3), 586.
Sartre, J. P. (1981). *Being and Nothingness*. London: Methuen and Co.
Scruton, R. (1994). *The Classical Vernacular*. Manchester: Carcanet.
Seamon D. (1979). *A Geography of the Life World*. London: Croom Helm.
Sherrington, C. S. (1973). *The Integrative Action of the Nervous System*. New York: Arno Press.
Thompson B., Hinks J. and Green P.(1998). *Escape Syntax*. In T.J. Shields (ed.). *Human Behaviour in Fire*, Fire SERT center. 799-808. Jordanstown: University of Ulster.
Thompson, W. J. (2001). An Evolutionary Approach to Spatial Knowledg. In Proceedings of the 3rd International Space Syntax Symposium, 50.1-50.10. Atlanta: Georgia Institute of Technology.
Thompson, W. J.(2003). Architectural Hermeneutics V. Harry and the Philosopher's Stone, *Environment and Behaviour*, Sage Publications, 35(4), 478-485.
Thompson, W. J. (1999). PhD Thesis Wayfinding in Complex Space. Herriot Watt University.
Tolman, E. C., Ritchie, B. F. and Kalish, D. (1946). Studies in Spatial Learning, *Journal of Experimental Psychology*, 36, 13-24.
Tolman, E. C. (1938). The Determiners of Behaviour at a Choice Point. *Psychological Review*, 45, 1-41.
Tooby, J. and Cosmides, L.(1992). The Psychological Foundations of Culture. In J. H. Barkow et al. (eds). *The Adapted Mind*. Oxford: Oxford Press.

Uexkull, J. Von. (1957). *A Stroll Through the Worlds of Animals and Men.* New York: International Universities Press Inc.
Wapner, S. et al. (1980). An Organismic Developmental Perspective for Understanding Transactions of Men and Environments. In G. Broadbent (ed) *Meaning and Behaviour in the Built Environment.*79-91. Chichester: John Wiley and Sons.
Wirth, L. (1964). *On Cities and Social Life.* Chicago: Phoenix.
Woodcock, G. (1946). *William Godwin.* London: Porcupine Press.
Young, R. M. 1994). *Mental Space.* London: Process Press Ltd.
Zeisel, J. (1984). *Inquiry by Design.* Cambridge: Cambridge University Press.
Zeleny, M. (ed) (1980). Autopoiesis, Dissapative Structures and Spontaneous Social Orders, AAAS Selected Symposia. Boulder Colorado: Westview Press.
Zimring, C. M. (1987). *Evaluation of Designed Environments.* New York: Van Nostrand.

Chapter 12

The Influence of Developmental Maturity in the Environmental Representation of the City: An Empirical Approach

Ángel Fernández González

The purpose of this research is the verification of the influence of maturity, understood herein as an evolutionary development associated with age in the representations or environmental images in the city within two groups of young students divided by an age span of 4 years (from 17 to 21 respectively) which maintains the variable 'origin' constant. All of the aforementioned subjects were born and have always lived in the same city.

The representations of the spatial environment evolve as a person develops, as Piaget and his collaborators' pioneer investigations have already demonstrated (cf. Piaget, 1966; Piaget and Inhelder, 1967). In fact, Piaget's work displays the first theoretical model which satisfactorily relates the development of spatial cognition to cognitive development, studying the development of the representation of space or the ontogenesis of spatial comprehension. We find other important developments of this theory in Hart and Moore (1973) and in Siegel and White (1975).

Nowadays, we know that age, in what concerns development or maturity, is a crucial factor in understanding spatial representations (Spencer, Blades and Morsley, 1989; Kitchin and Freundschuch, 2002; Liben and Downs, 2003; Quaiser-Pohl, Lehmann, and Eid, 2004), either because as time goes by one develops or uses different reference frameworks to spatially orientate himself or because one evolves in terms of the precision and complexity of the spatial representations that one has in his/her memory (cfr. Revisions: Acredolo, 1981; Evans, 1980; Hart and Moore, 1973; Heft and Wohlwill, 1987; Matthews, 1992; Moore, 1976; Siegel and White, 1975; Spencer, Blades and Morsley, 1989; Torell, 1990; Utal and Tan, 2002).

This is the reason why we can speak about two lines of research in this sense. One of them investigates, above all in children, the sequence or development of the reference frameworks which are used by people to orientate themselves in space by finding a sequence: first, by use of egocentric signs, then by stable objects in space (separated then coordinated) up to the point of using systems of reference both coordinated or independent of the final subject. This research offers results which are in accordance with Piaget's contributions (Piaget, 1966, Piaget and Inhelder; 1967; Piaget et al., 1960) and those of Hart and Moore (1973). However, that sequence cannot be considered as something rigid and inflexible, as Acredolo (1976, 1977) points out; if one provides adequate signs of reference or landmarks, the egocentric responses in young children can be reduced. Likewise, one must bear in mind the type of work that is given to the subject, the scale, etc. (Bell, 2002; Heft and Wohlwill, 1987; Heth and Alberts, 1997) so as to be in spatial works and so as not to underestimate their abilities (i.e. five-year-olds can read and understand vertical aerial photographs, which demonstrates their spatial abilities: cf. Plester et al., 2002).

On the other hand, in research about reference frameworks and orientation in older people, it became clear that these were less precise or accurate than in the case of the young adults in their orientation skills (Looft and Charles, 1971; Rubin et al., 1973), and they made more non-egocentric mistakes (Schultz and Hoyer, 1976), fewer correct judgements and a higher number of mistakes in identifying landmarks than the young adults did (Ohta et al., 1977) (cf. Evans, 1980; Kitchin and Freundschuch, 2002).

A second line of research, as we have already shown, was centred on the accuracy and complexity of spatial representations. Along this line we can notice a model like that of Siegel and White (1975), which outlines a development of the spatial representation according to which children perceive and remember space following this sequence: first landmarks, then paths and finally an entire reference framework, a sequence which implies an elaboration of the formulation by Piaget et al. (1960). This research, as well as that of Shemyakin (1962), indicates that at the minimum age of 9 or more a coordinated system of landmarks and paths of a holistic nature start to develop. With adolescents, establishing themselves in the accuracy of drawing a map, few changes are observed between 12 and 17-year-olds (Evans, 1980). It is also observed that the older subjects can display deficiencies in environmental cognition; the subjects over sixty manifested less knowledge of the environment than the adults over 26 (Barrash, J., 1994; Evans et al., 1984). Obviously, these differences attributable to the age could be explained by the quantity and quality of the moves made by the subject (Evans, 1980; Rissotto and Tonucci, 2002), as well as the fact that, among other things, the life cycle produces different environments which stand out in his/her mind (Michelson, 1977).

Curiously, we notice that the ontogenetic sequence in the environmental representation which derives from the model previously exposed also has its replica in the microgenetic development of adults, after arriving in a new environment and slowly but surely familiarizing themselves with it. First, they find out about the landmarks, then the paths (Garling et al., 1981; Evans and Pezdek, 1980; Evans et al., 1981), and once the system of paths is formed, is it easier for them to remember the whereabouts of the landmarks and points of reference. These people would be stuck to the origins and destinations of the paths (Golledge, 1987) so it would be logical to have a sequence of sign learning. Garling and Cols (1981) found that living in a medium-sized Swedish city, one could remember with great precision where the landmarks were and yet it was more difficult to remember the system of paths. All the same, the spatial abilities increased the speed of learning, although the distortions in the representation could remain in their memory for a long time.

On the other hand, in research with adults, we sometimes notice contradictory results (i.e. with respect to perception in the centre of the city: cfr. Aragonés, 1986), which show that in addition to the importance that time or age of learning can have, a principal factor which could explain the differences would be the familiarity or personal experience of the surroundings. A wider experience or familiarity of the surroundings exists at an older age or developmental stage. In any case, we should bear in mind all the nuances which were pointed out above and the fact that they can change the predictions from the outset.

In this context, the primary target of this work is to analyse the influence of the age factor (Secondary School vs. University) in the knowledge and appraisal of the city.

Concepts, Variables and Indicators

Subjects' knowledge of the city is operationalised by means of direct self-evaluations (subjective knowledge) and by indicators on a map-drawing task: number of elements sketched, number of elements identified and proportion of mistakes in identifying elements. The greater the number of elements correctly identified, and the lower the number of incorrectly identified elements, the greater the subject's knowledge of the city is assumed to be; this is what we have termed objective knowledge, in the sense that it is based on the quantifiable results of the drawing.

Appraisal of the city is operationalised by means of direct questions (level of satisfaction with the city, degree to which subjects like or dislike the city), and by means of a grid procedure, which was later assessed using a factorial analysis, revealing the typical Osgood semantic differential structure: evaluation, image (power), and activity.

Method

Subjects

In this study, we are working with a sample made up of students (n=113) from two levels of age and maturity: high school students (n=52) and university students (n=61). All of these share the fact that they were born and have always lived in the same city, Ourense, which has a population of slightly more than 100,000 inhabitants. When working with students who have always lived in the city, and ignoring others, we intended to control the effects of familiarity which could arise due to the different origin of the students. This is the reason for maintaining the constancy of the said variable for its control.

We must take into account, however, that this type of sample is not experimental in the strict sense of the word, as uncontrollable factors may come into play, especially those related to the tendency to continue one's university studies in his/her own city. This influence cannot be denied and therefore must be taken into account when interpreting the results. Nevertheless, in this particular case, in which the majority of students continue to study up to the university level, we do not consider this factor to be decisive. Even so, we must remember that this is a *post hoc* study, and as such any conclusion we draw must be considered tentative.

The average age of the high school students is around 17 years old (16.91), as opposed to those of 21 (20.93) years of age from the university sample. The subjects were chosen at random from previously made up groups (classrooms) belonging to two institutions where high school education is taught (Otero Pedrayo and As Lagoas) and three university departments (Education, Humanities and Science).

Instruments

Three different tests were applied to gather information (a sketch map or a map with a drawing of the city; a questionnaire about the subjects' mental representation of the city, and a semantic differential applied to the city) with the aim of externalizing the image and representation of the city in question.

The drawing of a map consisted in having the subject draw a map of the city. The questionnaire of representation of the city consisted in open and closed questions regarding their knowledge of the city (level of familiarity with the city, ease with which they would be able to find an address, etc.), representation (what comes to their heads when they think of the city, etc.), and appraisal of the city (to what degree they like the city, are satisfied living there, etc.). The semantic differential is composed of 23 scales or bipolar adjectives related to the city.

Procedures

The application of the tests was carried out collectively, as is usually done with this type of population, as they are made up of natural groups. It took about 55 minutes to carry out the three tests.

As for the task, the subjects were asked to draw a map of the city on a blank sheet of A4 paper in such a way that a person who did not know the city would be able to have an idea of what it was like. The maximum time allotted for this task was 25 minutes. The subjects were also asked to order the elements on the map in the order in which they drew them. Then, upon finishing the drawing, the subjects identified the elements they were able to from their drawing and wrote them down on another sheet of paper. The second task involved answering some open-ended questions as quickly as possible (i.e. what came to mind when they thought of the city, which places they liked best in the city, etc.) and closed questions (about their knowledge and appraisal of the city, choosing from among 4 alternatives: a lot, quite a bit, very little, not at all). For the third and final task, a semantic differential was applied in regards to the city. Each subject was asked to choose from among 7 options along a bipolar scale in such a way that it reflected his/her appraisal of the city, having been given an example beforehand to aid them in completing the task.

Results

Once the externalisation of the image or mental representation of the space of the city has been obtained, it is a matter of finding out if there are significant differences between these two groups of students from the high school and university samples with a different level of developmental maturity associated with age.

A first approach to the results allows us to differentiate between subjective knowledge (i.e. what the subjects say they know about the city) and objective knowledge (i.e. what we deduce they know on the basis of the applied tests). The differences between the two groups of students regarding the level of maturity or development - high school versus university students - have to be analysed using as a reference these two types of groups in question. In this way, both groups, differentiated regarding the level of maturity, claim they have similar subjective knowledge and do not show significant differences in this aspect when measured, and yet we know, with reference to objective knowledge, that there are significant differences between both groups (because the older group, superior in maturity and development, is superior in that knowledge).

In this project, using age as a reference, we want to establish a comparative analysis concerning the maturity of the subjects. We are going to contrast the representation of the city by means of different indicators in the two groups of students in question.

We will start by comparing what we call objective knowledge (which was obtained from different aspects of the knowledge which the subjects displayed in the drawing they did of the city) and subjective knowledge (what the subjects said they knew) in the two groups in question (students from high school and universities). The differences between the groups are also analysed (see Table 12.1).

Table 12.1 Knowledge and appraisal of the city in terms of maturity

Variables	Secondary	University	$F_{(1,109)}$	Discriminant structure
	Media	Media		
Drawn elements	20.19	26.11	8.374**	0.549
Identified elements	14.83	20.79	14.887**	0.721
Proportion of overall mistakes	0.6171	0.5100	1.491	-0.286
General satisfaction	3.7885	3.7541	0.122	-0.030
Subjective knowledge	3.3769	3.4295	2.436	0.087
Appraisal	4.6462	4.2459	8.293**	-0.470
Image	4.4269	4.1115	2,436	-0.243
Activity	3.8692	3.7000	0.241	-0.165
Effects in term of age: $\lambda = 0.773$ $\chi^2 = 27.529$ g.l. = 8 $p(\chi^2) = 0.001$ * $p < 0.05$ ** $p < 0.01$				
Box test: F (36, 39321) = 0.756 p (F) = 0.853 M = 29.506				

As we can see, some significant differences exist globally between both groups of age and maturity (F4.123 p(F)<0.05). This global difference is stated explicitly in variables or indicators related to what we call objective knowledge or the knowledge the subjects manifested through the number of drawn elements in the pictures they did of the city (F 8.374; *p*<0.01), whereas there was no sign of significant differences between the groups in what we call subjective knowledge or what students from both groups said they knew about the city (F 2.436; ns). This is reflected in the levels of probability as well as in the discriminant structure. This type of variable allows for the correct classification of 70.8% of the cases examined.

Similarly, a significant difference was found in one of the three factors, particularly in the appraisal of the city (what if offers), which we find as a result of applying the factorial analysis to the scales of the semantic differential of the city. This analysis allows us to obtain three factors (see Table 12.2), which are used here as variables of contrast (appraisal of what the city has to offer, image of the city, activity in the city). An F 8.293 ($p<0.01$) is observed in the appraisal of what the city has to offer, which shows us that here lies another of the indicators of the existing differences in the representation of both comparison groups.

Table 12.2 Matrix of inverted components

	Component		
	Appraisal	Image	Activity
Services	0,603		
Interesting	0,597	0,317	
Green areas	0,568	-0,136	0,168
Populated	-0,513	0,407	0,378
Well designed	0,496	0,347	0,154
Modern	-0,480		
Traffic	-0,403	0,143	0,391
I know it	0,297	0,144	0,177
Pleasant	0,532	0,618	
Desire to live in	0,137	0,605	
Economy		0,592	
I like it	0,522	0,589	-0,104
Important		0,514	0,134
Clean		0,388	
Popular	0,222	0,339	
Good access	0,198	0,316	
Active	0,168	0,151	0,709
Quiet		-0,223	0,709
Rich			0,685
Cheap	-0,166	0,139	0,642
Shops	0,422	0,348	0,448
Safe		0,373	-0,439
Reputation	0,368		0,414

Method of extraction: Analysis of principal components.
Method of rotation: Normalisation Varimax with Kaiser.
The rotation converged in 5 interactions.

To greater appreciate the tendencies in the expression of these differences, we can make use of the average scores which have been written down for both groups in each one of the variables used as an element of comparison. If we refer to the last variable we mentioned, in which there were significant differences (appraisal of the city), we can verify in Table 12.1 that the average appraisal of the Secondary School students is greater, though their knowledge is surprisingly lower than that of the University students with regards to the city. This curious observation could suggest a more idealised image of the city among the younger participants, due to their lower degree of maturity.

If we would like to see a more detailed semantic appraisal of the city comparing the two groups studied here, we can observe in Table 12.3 that the significant difference in the factor of appraisal, shown above, (appraisal of the city, in Table 12.2) is explicitly stated in the scales related to the urban design of the city and the importance given to it. In these scales, one can note some significant differences between groups which in the same way remain reflected in the discriminant structure, showing us where the discrimination between groups is maximised.

Table 12.3 A semantic appraisal of the city among students, in terms of age and maturity (secondary school vs. university)

Variables:	Secondary School	University	Discriminant function structure
Ease	3.27273	3.36667	-0.052
Interesting	4.61818	4.08333	0.259
Well-designed	3.81818	3.13333	0.347
Rich	4.23636	3.80000	0.275
Reputation	4.78182	4.53330	0.143
Active	4.38182	3.80000	0.309
Green areas	2.94545	2.86667	0.041
Cheap	4.89091	4.66667	0.131
Public services	4.12727	3.78333	0.171
Modern	3.85455	3.36667	0.268
Pleasant	4.92727	4.58333	0.201
Important	4.01818	3.13333	0.470
Safe	4.98182	4.65000	0.235
Populated	3.67273	3.86667	-0.114
Popular	3.21818	2.98333	0.129
Good shops	5.12727	4.61667	0.275
Good access	4.63636	4.65000	-0.007
I like it	4.94545	4.85000	0.042

Clean	3.60000	3.91667	-0.171
Economy	3.78182	3.26667	0.288
Desire to live in	4.81818	4.56667	0.123
I know it	5.43636	5.41667	0.011
Traffic	4.63636	4.50000	0.080
Effects in term of:			
Well-designed	F 4.736	p(F) < 0.05	
Important	F: 8.697	p(F) < 0.05	
Box test:	F: 1.230	Sig. 0.006	

If we study the variables of knowledge thoroughly, we can see that the situation clearly changes. It is the university students who show superior objective knowledge, following our terminology, as they are the ones who clearly draw more elements of the city, as the average scores show, and they are also the ones who identify more places on the drawings. This higher level of knowledge suits the higher level of maturity of this group.

The most obvious difference between the two groups is that the older participants drew more complete maps. In any event, by combining the information obtained (i.e. more elements drawn, a greater number of elements identified, fewer errors), we can draw the overall conclusion that the University students have a greater degree of objective knowledge of the city than their secondary school counterparts, as measured using standardised, external assessment criteria rather than self-perceptions.

Discussion

In general, the results offer evidence as to the existence of differences in knowledge and appraisal in the sample studied.

Everything that we have said before about the average scores, with respect to the tendency toward differences in the objective knowledge of the city, has to be considered bearing in mind that the subjects of both groups claim to have a similar level of subjective knowledge. In the same way, the percentage of mistakes which they make when they do a drawing is similar, or much the same and no significant differences exist between both comparison groups.

Nonetheless, it is important to keep several things in mind. First of all, the comments made before when speaking about the sample must be considered. Secondly, subjects' artistic abilities may have some influence on their drawings of the city. Thirdly, while it is true that subjects' memory of relative location increases with age, the way in which they interact with space also affects the point of reference they use and as such their representation of the environment (Bell, 2002). In addition, different formats for mental representations of the environment are used depending on the way in which subjects memorise daily or routine journeys around their city in a chronological sequence (Helstrup and Magnussen, 2001). Moreover, we must consider individual differences in terms of spatial abilities (spatial visualisation, orientation, and relationships), which may affect one's performance on tasks involving orientation (Malinowski and Gillespie, 2001).

While there are studies which indicate that age is the most important predictor of spatial knowledge (Bourchier, Barrett and Lyons, 2002), it is important to bear in mind that the limitations placed on children's and young people's autonomy may affect their acquisition of environmental knowledge, as empirical data confirms the importance of direct interaction between an individual and his/her environment in the development of such knowledge.

In conclusion, after taking into account all of the aforementioned considerations and analysing the data obtained, we have found various differences between the two groups studied. The older, more mature group appears to be more familiar with the city and to have a more critical attitude toward it. The younger group, however, while claiming to have a similar level of familiarity with the city as in the older group, showed themselves to be less knowledgeable than their older counterparts when asked to carry out a practical demonstration of that knowledge. In addition, the younger subjects were found to have a significantly more positive attitude toward the city than the older participants.

To sum up, in the sample studied, one can see that the level of maturity of the subjects is related to their knowledge as well as their appraisal of the city, directly in the first case, and inversely in the latter. This is probably due to the fact that as young people mature, they develop a less idealistic, more realistic view of the city.

References

Acredolo, L.P. (1976). Frames of Reference Used by Children for Orientation in Unifamiliar Sapaces. In G.T. Moore and R.G. Golledge (eds.). *Environmental Knowing,* (pp. 165-172). Stroudsburg, Pa.: Dowden, Hutchinson and Ross.

Acredolo, L.P. (1977). Developmental changes in the ability to coordinate perspectives of large-scale environment. *Developmental Psychology.* 13, 1-8.

Acredolo, L.P. (1981). Small-and Large-Scale Spatial Concepts in Infancy and Childhood. In L.S. Liben, et al. (eds.). *Spatial Represenation and Behavior Across the life span,* (pp. 63-82). New York: Academic Press.

Aragonés, J.I. (1986). Cognición ambiental. In F. Jiménez Burillo and J.I. Aragonés (eds.). *Introducción a la Psicología Ambiental,* (pp. 66-83). Madrid: Alianza.

Barrash, J. (1994). Age-related decline in route learning ability. *Developmental Neuropsychology.* 10, 189-201.

Bell, S. (2002). Spatial cognition and scale: a child's perspective. *Journal of Environmental Psychology.* 22, 9-27.

Bourchier, A. Barrett, M. and Lyons, E. (2002). The predictors of children's geographical knowledge in elementary school children. *Journal of Environmental Psychology.* 22, 79-94.

Evans, G.W. (1980). Environmental Cognition. *Psychological Bulletin.* 88 (2), 259-287.

Evans, G.W., and Pezdek, K. (1980). Cognitive mapping: Knowledge of real-world distance and location information. *Journal of Experimental Psychology: Human Learning and Memory*. 6, 13-24.

Evans, G.W., Marrero, D. and Butler, P. (1981). Environmental learning and cognitive mapping. *Environment and behavior*. 13, 83-104.

Evans, G.W., Brennan, P.L., Skorpanich, M.A. and Held, D. (1984). Cognitive mapping and elderly adults: verbal and location memory for urban landmarks. *Journal of Gerontology*, 39(4), 452-457.

Gärling, T., Böök, A., Erguezen, N., and Lindberg, E. (1981). Memory for the spatial layout of the everyday physical environment: Factors affecting the rate of acquisition. *Journal of Experimental Psychology*. 1, 263-277.

Golledge, R.C. (1987). Environmental Cognition. In D. Stokols and I. Altman (eds.), *Handbook of Environmental Psychology*.(Vol. I, pp. 131-174). New York: John Willey & Sons.

Hart, R.A. and Moore, G.T. (1973). The Development of Spatial Cognition: A Review. In R.M. Downs and D. Stea (eds.), *Image and Environment*, (pp. 246-286). London: Arnold.

Heft, H. and Wohlwill, J.F. (1987). Environmental cognition in children. In D. Stokols and I. Altman (eds.), *Handbook of Environmental Psychology*, (Vol. I, pp. 175-203). New York: John Wiley and Sons.

Helstrup, T. and Magnussen, S. (2001). The mental representation of familiar, long-distance journeys. *Journal of Environmental Psychology*. 21, 411-421.

Heth, C.D. and Alberts, D.M. (1997). Differential use of landmarks by 8- and 12-year-old children during route reversal navigation. *Journal of Environmental Psychology*. 17, 199-213.

Kitchin, R. and Freundschuch, S. (2002). *Cognitive mapping: past, present and future*. London: Routledge.

Liben, L.S. and Downs, R.M. (2003). Investigating and facilitating children's graphic, geographic, and spatial development: an illustration of Rodney R. Cocking's Legacy. *Journal of Applied Delevelompental Psychology*. 24(6), 663-679.

Looft, W. and Charles, d. (1971). Egocentrism and social interaction in young and old adults. *Aging and Human Development*. 2, 21-28.

Malinowski, J.C. and Gillespie, W.T. (2001). Individual differences in performance on large-scale, real-world wayfinding task. *Journal of Environmental Psychology*, 21, 73-82.

Matthews, M.H. (1992). *Making sense of place: Children's understanding of large-scale environments*. Harvester Wheatsheaf: Barnes and Noble Books.

Michelson, W. (1977). From congruence to antecedent conditions: A search for the basis of environmental improvement. In D. Stokols (ed.), *Perspectives on Environment and Behavior*. (pp. 205-220). New York: Plenum Press.

Moore, G.T. (1976) Theory and research on the development of environmental knowing. In G.T. Moore and R.G. Golledge (eds). *Theories, research, and methods*. (pp. 138-164). Stroudsburg, PA: Dowden, Hutchinson and Ross.

Ohta, R., Walsh, D. and Krauss, I. (1977). Spatial perspective-taking ability in young and elederly adults. Paper presented at the 85th. Annual meeting of A.P.A.: San Francisco, August 26-30.

Piaget, J. (1966). *The Psychology of Intelligence*. New York: Littlefield, Adams.

Piaget, J. and Inhelder, B. (1967). The child's conception of space. New York: Italy.

Piaget, J., Inhelder, B. and Szeminska, A. (1960). *The child's conception of geometry*. New York: Basic Books.

Plester, B., Richards, J., Blades, M. and Spencer, C. (2002). Young children's ability to use aerial photographs as maps. *Journal of environmental Psychology.* 22, 29-47.

Quaiser-Pohl, C., Lehmann, W. and Eid, M. (2004). The relationship between spatial abilities and representations of large-scale space in children - a structural equation modeling analysis. *Personality and Individual Differences,* 36 (1), 95-107.

Rissotto, A. and Tonucci, F. (2002). Freedom of movement and environmental knowledge in elementary school children. *Journal of Environmental Psychology.* 22, 65-77.

Rubin, K., Attewell, M. and Tumolo, P. (1973). Development of spatial egocentrism and conservation across the life span. *Developmental Psychology,* 9, 432.

Schultz, N. and Hoyer, W. (1976). Feedback effects on spatial egocentrism in old age. *Journal of Gerontolog,* 31, 72-75.

Shemyakin, F.N. (1962). General problems of orientation in space and space representations. In B. G. Ananyev (ed.). *Psychological science in the USSR.* 186-251. Washington, DC: U.S. Office of Technical Reports.

Siegel, A.W., and White, S.H. (1975). The development of spatial representations of large-scale environments. In H.W. Russell (ed.), *Advances in Child Development and Behavior.* (pp. 10-48). New York: Academic Press.

Spencer, C., Blades, M. and Morsley, K. (1989). *The child in the physical environment: The development of spatial knowledge and cognition.* Oxford: John Wiley and Sons.

Torell,-G. (1990*). The acquisition and development of environmental cognition in children.* Sweden: University of Goeteborg.

Utal, D. and Tan, L.S. (2002). Cognitive mapping in childhood. In R. Kitchin and S. Freundschuch. *Cognitive mapping: past, present and future.* London: Routledge.

Chapter 13

The Home as a Territorial System (II)[1]

Mariann Märtsin and Toomas Niit

Theoretical Background

Environment as the Regulator of Individual's Openness/Closedness

An individual's being is not merely a physical body possessing a soul, rather, it is that the individual's realm has expanded to encompass the surrounding environment - in this sense; we can speak about the *environmental person* (Heidmets, 1988). Heidmets, who points out that this type of expansion involves both the social and mental, as well as the physical environment, calls the expansion of self into the surrounding environment, the personalization of environment. Personalization of the environment or the extension of the self into the world around us is governed by the need for regulating the relationships between people and bringing order into the world (*Ibid.*). Through the personalization of the environment, people take control over the external world, a part of it belongs to the person, s/he controls the things happening inside that part and s/he acts as a subject in relation to that part of the world. This way, a person decreases his/her dependence from the external world and makes at least some part of it predictable. From the other perspective, personalization enables an individual to regulate his/her openness/closedness in relation to the external world. As the personalized part of the external world acts as an extension of the self, of which a person has control over, an individual can decide to whom and what to tell about him/herself through the environment, s/he can decide when to open him/herself up to others and when to hide from their view.

The most frequently mentioned concept in the field of openness/closedness regulation is probably Altman's interpretation of privacy. Altman (1976) defines privacy as the selective control over access to the individual or group, seeing it as a boundary regulation process. The boundaries between the self and the non-self are permeable where their permeability varies across time and situations. That way, in regard to him/herself a person can regulate when and what information shall be given to the external world, and what kind of information will be received. This allows the person to gain the preferred level of openness/closedness.

---

[1] We apologize for borrowing the title from Vittoria Giuliani's (1988) paper, but we know that she does not mind. We are hopeful that there will be some interesting parallels found.

Altman points out that in the process of gaining the optimal level of privacy, a person may use various means: verbal mechanisms (content of the message, tone of voice, choice of words, etc.), nonverbal mechanisms (eye contact, body orientation, etc.), environmental mechanisms (particular elements of the physical environment, like doors and windows, and also personal space, distance, territoriality, etc.), and cultural norms and customs (when and what mechanisms to use) (Altman, 1976).

Pointing out that a person can regulate his/her openness/closedness with the help of environmental mechanisms, Altman (1976) states that the selectively permeable boundaries between self and non-self do not run along the body surface of the person but are further away in the environment. An individual has built him/herself outside of his/her physical body by expanding him/herself to the social, mental, and physical world. Although all objects of the social, mental, and physical world can be what provides personal information, this discussion concentrates on only one of the objects of the physical world, the home.

The Home

For a person, the home is probably the most important object of the physical world. The relationships between the person and the home are described by special cognitive, emotional, and behavioral ties, which make it possible for the home to be the regulatory means of the openness/closedness of the person. The person designs and decorates his/her home while taking into account his/her personal wishes and preferences. A lot can be learned from paying attention to the memorabilia, photos, etc. that are evident in the person's home. Visibility of the books, sports equipment, etc. give us hints about a person's hobbies and interests. The existence and visibility of different things in the home are very informative, but a lot can also be learned from those things that are missing or hidden in the home. Closing doors or leaving them open, erecting walls or breaking them down, and reserving a special room for receiving guests inside the home can also be seen as signs of openness/closedness regulation. All these indications give us information about the people living in the home and their private or open territories within the home.

In addition to the fact that the home provides the outside world with information about the people living in it, the home can also give information about the socio-cultural norms, customs, and attitudes accepted by the people living there. An example that illustrates how the socio-cultural norms are reflected in the design and decoration of the home would be the differences pointed out between a traditional Japanese home, and a modern American home. We cannot speak of private rooms for different family members that are closed to the outside world, and sometimes also to other family members in the case of a traditional Japanese home. Rather, we see that the space inside the home is divided and smaller spaces are created for different activities by movable walls (Omata, 1992). In contrast, the American home consists of different private rooms for individual family members and in addition to that, there are separate rooms used for receiving guests and for family gatherings (Altman and Chemers, 1980). These kinds of differences in room usage can be seen as a reflection of different cultural contexts - as for the collectivistic culture of the Japanese, it would not be suitable to separate family members into different private rooms, whereas, in the American home, the interior design clearly reflects the individualism stressed in that cultural background (Märtsin, 1999).

In conclusion, the home can be seen as a means for providing the outside world with information about the people living in it, and at the same time it is a reflection of the wider socio-cultural norms and customs that have been accepted by those people. In this way, the home draws the line between the self and non-self, and provides the individual with the opportunity to separate oneself from the external world and at the same time; it also ties the person into the wider social context.

The Home as the Regulator of the Family's Openness/Closedness

Thus far, an overview of the regulation of openness/closedness of the individual through the environment, which includes the home, has been given. It is usual in our cultural context that the home is a place for living, not only for one person but also for the people who are connected to each other as a family.

Altman's model of privacy, which is the basis for the current discussion, does not consider only the individual's regulation of privacy, but can also be used to describe similar processes at the group level (Altman, 1976). Considering the individual, the boundary that needs to be regulated is drawn between the *self* and *non-self*, in the case of a group, the differentiation is made between *us* and *them*. Inside the group boundaries, are the people who share similar understandings, attitudes, value orientations, etc., and outside, are the people whose characteristics are different from those of the particular group. It is also the case with groups that the boundaries do not serve only a function of separation, but here as well different means are used to give others information about the group and thereby provide the group with information about the outside world that could be accepted and taken into account. The group also expands itself into the external environment, using its objects as privacy regulation mechanisms.

Family, as a social group can also use different *environmental objects* as privacy regulation mechanisms. One environmental object that can be used as a privacy regulation mechanism is the home. In the previous discussion we gave examples of the way the home can express the self of an individual and how it can also express the information the individual has accepted from the world around him/herself. Similarly, as the home expresses the identity of the family, it also demonstrates who the people are that are considered part of the family and gives information about the activities these people are involved in. Paying attention to room usage at home, to who it is that certain rooms belong to, who has a private room, and who has to share a room with another person gives us important information about the relationships within the family. Room usage at home can also give us information about the wider socio-cultural norms accepted by the family - whether or not the father has his own study, or is the extra room used for providing every child with his/her own room. In this way, the home also functions as a mechanism for connecting the family to the outside world. Using different rooms for receiving guests and preparing to show hospitality prior to reception, can also be seen as signs of this type of connection process.

By expressing the identity of the family as a whole and regulating its openness/closedness, the home still does not lose its function as a privacy-regulating mechanism for different individuals living in the home. The modern American home can once again be an example of this process. There are rooms inside the American home used by all members of the family, which express the identity of the family as a whole and that can be used for receiving the family's guests. At the same time, there are private rooms for family members that allow expression of self for certain family members, where the family's common guests are seldom received (Altman and Chemers, 1980).

In light of the previous discussion, it seems justified to state that the home regulates the openness/closedness of people living in it on two different levels: from one perspective it draws a line between the family and the external world, and from the other perspective, it functions as a privacy regulator for separate individuals. In making such a hypothesis, it is necessary to take into account the cultural context in which the family is living since most probably the home functions as a privacy regulation mechanism for the whole family in all cultural contexts, but stands to regulate the privacy of individual family members only in more individualistic cultures. The differences pointed out between the traditional Japanese home and American home could once again be used as examples (Canter, 1977; Altman and Chemers, 1980).

The Home as a Territorial System

In the present study the home is seen as a unified whole that provides individual family members and the family as a whole with the opportunity to be separated from the external world; however, at the same time, the home functions as a connection mechanism to the outside world for the family as a whole, and its individual members. It seems that not all the parts of the home have these opposite functions working at the same time; rather, certain differentiation in space and time should be made for these functions inside the home. Taking that into account, it seems justified to view the home as a system consisting of different territorial units, which as a unified system provides the family and individual family members the opportunity for regulating their openness/closedness towards the external world.

Normally, the distribution of activities between different parts of the home have been viewed as foundational in defining different territorial units inside the home (e.g. family member's bedrooms as private areas of the home and guestroom as a public zone in the home). Still Sebba and Churchman (1983) point out that human territoriality can not be considered only in behavioral terms, like defending the place or controlling it, but it also covers topics like belonging, owning, personalization, etc. (e.g. Edney, 1976; Brown, 1987). Therefore, the attitudinal aspect in addition to the behavioral should also be taken into account in differentiating territorial units inside the home. The authors stress that in considering different activities we should also pay attention to potential activities in a certain area (e.g., How would a person behave towards other people in that area if that person would like to be alone?) and activities in relation to that area (e.g., Is that person engaged in the maintenance of that area?), in addition to studying the areas where different activities are held. Furthermore, we should also try to find out individual family members' attitudes toward different parts of the home - belonging, expressing the self, etc.

Other authors have followed Sebba and Churchman's study (1983) on territorial units inside the home as well. Just to mention some of these: Omata (1995) has studied Japanese home-wives' territoriality and its relations to the room usage at home and psychological well-being; Giuliani, Rullo, and Bove (1990) have studied Italian family's room usage in relation to eating and sleeping activities; Rullo (1992a, b) has studied differences between room usage, territoriality behavior, and satisfaction with the home in the case of young adults living in their parent's home and young adults living separately.

A review of the literature reveals that there are different theoretical concepts and empirical studies that consider the territorial units of the home. Summing up the findings of those indicates that there are three territorially different units inside the home – (a) a public room meant for receiving guests; (b) a family room meant for family gatherings; (c) private rooms for individual family members (Omata, 1992; Altman and Chemers, 1980; Giuliani, 1988).

The Empirical Study

Goals and Hypotheses

Proceeding from the aforementioned theoretical framework, the empirical research plan was put together with the goal of finding out whether it is possible to distinguish different territorial units within the home, and how these units can be what categorized. Another goal was to figure out whether it is possible to interpret the results proceeding from the concepts of openness/closedness presented in the theoretical part. The specific research hypotheses that were proposed are as follows:

1. Proceeding from the location of activities, we can distinguish two territorial parts inside the home – public activities are located in some room(s) (living room, guest room, etc.), and private activities are located in other rooms (family members' bedrooms, etc.);
2. In addition to the public/private distinction, we can also draw the distinction between the room(s) meant for the family and the part of home meant for receiving guests;
3. Receiving personal guests may take place both in the room where the family unit receives their guests, as well as in the part of the home where an individual family member considers the area as belonging to him/herself only;

4. For spending time alone different members of the family usually retire to their bedrooms;
5. There may be some overlap in the case of rooms that different family members consider belonging to themselves, which they consider as their personalized space, separated from others, and closed to others;
6. The rooms, where particular family members sleep, are the rooms they consider belonging to themselves, which they consider as their personalized space, separated from others, and closed to others.

Using a similar study as an example (Sebba and Churchman, 1983, this study used the spatial location of different activities as a starting point. In addition to that, attitudes of different family members toward different parts of the home were also studied. This approach also relies on our earlier conceptualizations (cf. Niit, 1988), which considered attitudes and activities as important aspects in analyzing socio-physical systems at the individual level (Figure 13.1) (for explanations about the different arrows in this figure, cf. Niit, 1988, pp. 384-386).

Method

Sample

The family was defined as a legally married or cohabiting couple and their children. The sample was a 'snowball sample', where 54 families participated in the study. 16% of them did not have children, 25% one child, 50% two children, 6% three children, and 2% four children. 32 children answered the Child's Questionnaire, 56% of them were the oldest children in the family, and 44% were younger children in the family.

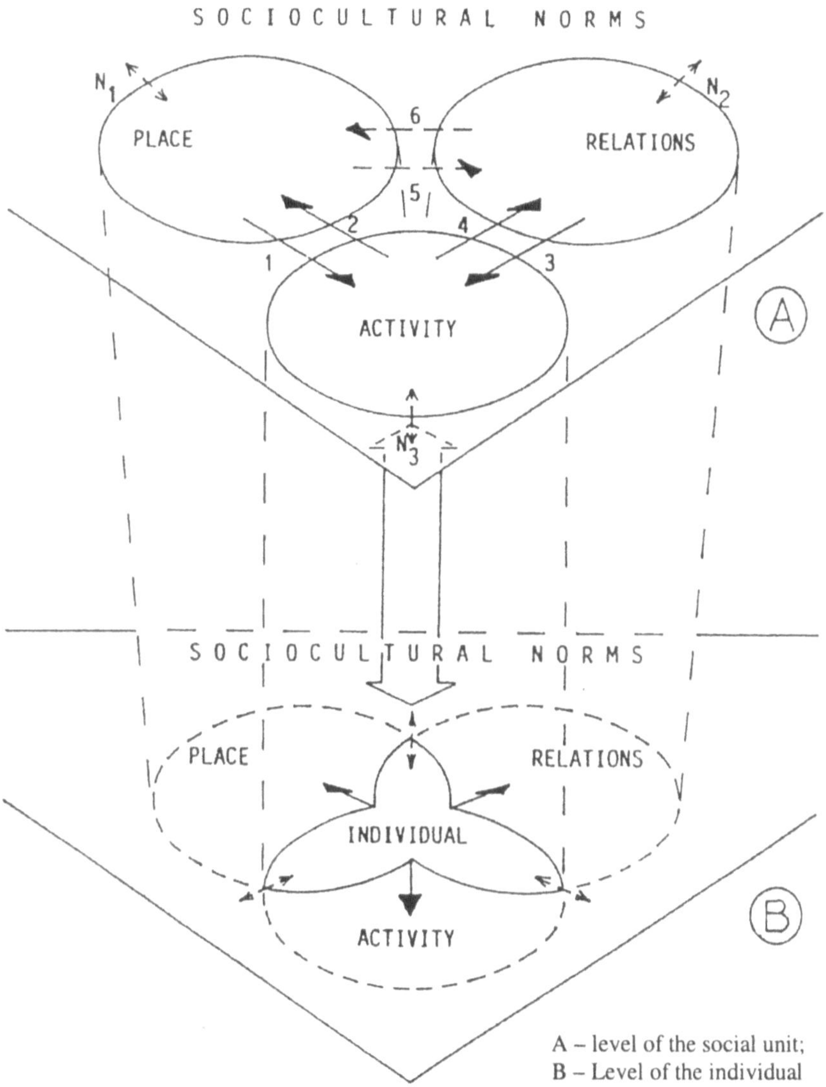

Figure 13.1 A framework for analyzing transactions between people and environment

Procedure

Special self-report questionnaires designed for this study were administered to the respective family members – *Wife's Questionnaire*, *Husband's Questionnaire*, and

Child's Questionnaire. All children in the family older than 10 years answered the latter one. All questionnaires consisted of four parts:

1. *General data about family and housing unit*. Questions about the family were included in Wife's Questionnaire and questions about the house or apartment in Husband's Questionnaire.
2. *Satisfaction of family members with their residence and living district*. These questions were presented to all family members.
3. *Attitudes of family members toward different parts of the home*. This part consisted of questions that ask about rooms' levels of privacy, and if they were secluded or restricted from others. They were included in all questionnaires.
4. *Activities taking place at home*. These activities were divided into two groups according to how disturbing the presence of others during a given activity is: public activities – receiving family guests, eating, and watching TV; private activities – sleeping, being alone, professional work or activities, and studying. In the case of children, playing was also considered, which could not be easily classified under the categories mentioned above. Although additional questions were presented for some activities, specific questions were asked for each case; where does this activity takes place most frequently inside the home, which activities can take place in this room simultaneously, and are these other activities disturbing for the respondent. The last two questions were proposed to all respondents about all activities (except 'receiving guests' and 'being alone'). Questions about spatial locations of the activities were asked about receiving guests, eating, and sleeping to female respondents; and watching TV, professional work, and children's play to male respondents; and about studying – to children. Questions about 'being alone' were asked to all family members.

Analysis

In addition to frequency distributions, cross-tabulations and comparisons of means, hierarchical clustering with Ward's method (1-Pearson's r) was used to describe the localization of activities inside the apartment (house).

Description of the Housing Conditions

61% of the families lived in an apartment they owned, 26% in their own houses, 4% in (mostly 2-storey) row-houses, and 9% in rented apartments. Almost half of the families lived in the 'stony' (i.e., high-rise) areas of Tallinn (Mustamäe, Õismäe, Lasnamäe, and the centre of the city).

The number of rooms per family was the following:

No. of rooms	**1**	**2**	**3**	**4**	**5**	**6<**
%	9	13	37	20	9	12

The average floor area was about 96 m^2 per family, 61% of the families lived on area less than 100 m^2. The average inside density (number of persons per room) was 1.07 (SD = 0.43), in case of 35.2% of the families there was more than one person per room. In average, these families had lived in their present residence 9.8 years (with minimum time of 4 months and maximum of 40 years). 59% of these families had not made any floor plan changes in their apartment or house.

Naming the Rooms

The living room (or 'big room') is clearly distinguishable from the other rooms. All other rooms are usually called 'bedrooms'. If the residence has more rooms, the additional rooms, other than the living room and bedrooms are given names like 'workroom' or 'computer room'.

Satisfaction of Family Members with the Residence and Residential Area

The respondents participating in this study were generally satisfied with the size of their residence, its layout, and residence in general (the average ratings for size were $X_{size} = 2.14$, SD = .66; for layout $X_{layout} = 2.17$, SD = .57; and for satisfaction with residence in general $X_{satisf} = 2.14$, SD = .58, on a 3-point scale). Statistically significant difference in satisfaction with the size of their residence was found between families with high and low inside density[2] ($X_{low} = 2.36$; $X_{high} = 1.75$, $p<.01$) and with their residence in general ($X_{low} = 2.26$; $X_{high} = 1.92$, $p<.01$).

35% of the respondents found that they would not need more rooms and 33% found that they would need one more room. 25% of the respondents thought that this room would be needed for the child and 3% mentioned that every family member would need his/her own room. In answer to the question why they would need more rooms, the most frequent answer was 'for the child', which was a common answer even among families who did not have a child at the time of response. In addition to that, it was said that a room for professional work would be useful (it was often referred to as *office* or *computer room*), as well as a parent's bedroom. Less often mentioned was the desire for a guestroom, a room for household work (sewing, ironing, etc.), or a wardrobe.

Generally, the participants of the study were satisfied with their place of residence ($X_{town} = 2.45$ and $X_{neighborhood} = 2.16$), although a statistically significant difference was found between children and parents in satisfaction with the city ($X_{children} = 2.65$; $X_{parents} = 2.39$; $p<.05$). In the questionnaire, 59% of the respondents had mentioned the area of the city to where they would like to move. Most of the areas mentioned were the 'greener' parts of Tallinn – in addition to Nõmme (14%) and Pirita (7%), several areas nearby the center were mentioned. Of the respondents, 11% would like to move out of Tallinn and 7% to the areas nearby.

---

[2] The group with high density were nineteen families involved in the study, who had more than one person per room.

The Attitudes of Family Members toward Different Parts of the Home

The Space Belonging to a Family Member

Of the 140 respondents who answered the questionnaire, 69% declared that in their home there are rooms or parts of rooms that belong specifically to some member of the family, whereas 24% said that there are no such rooms in their home. As the study described, all respondents were asked to describe the rooms belonging to them, and additionally mention the rooms belonging to the other family members. As such, it was possible to compare what rooms the respondents consider as 'theirs', as well as which rooms are recognized as 'theirs' by other family members. Most often, the parents' bedroom (36%), children's room (28%), and living room (14%) were mentioned by family members as being 'their rooms'.

If we look at the answers of the family members separately, then the parents' bedroom was considered as belonging to them by husbands and wives (100%), and children's room was considered as 'theirs' by children (96%). If the bedroom was considered belonging equally to husband and wives, the room referred to as the study was considered as belonging to the men only, and the kitchen was mentioned as belonging to women only.

The understanding we derive from other family members' evaluations is rather similar: 98% who considered the bedroom belonging to somebody, considered it belonging to parents, and all parents considered the children's room belonging to the children. The living room was usually considered to belong more to parents than to children. All the people who considered the kitchen as belonging to somebody, found that it belonged to the wife and that the study room belonged to the husband.

Personalized Space

Of the respondents, 44% expressed the opinion that there is no room or part of a room that allows expression of their personality; however, 36% said that there is such a space in their home. The living room was mentioned most often as a personalized part of space (17%), which was followed by the kitchen (16%), parents' bedroom (14%), and children's room (14%). Of the respondents, 13% mentioned that all the rooms in their home express their personality.

In looking at the answers separately, it is evident that the parents consider their bedroom as a personalized room (husbands and wives consisted of 83% of these who mentioned parents' bedroom as a personalized room), and 93% of the children considered their room as personalized.

Only husbands considered their study as being a room that expresses their personality; whereas, the majority who mentioned the kitchen as this room was women. The living room (or parlor) was considered mostly by women to express their personality (57% of the wives mentioned the living room as a personalized space).

Private Room

In this study, there was more people who have a room where they can be left undisturbed if they wish, as compared to those who did not have such an opportunity. Of the respondents, 38% mentioned the parents' bedroom as this type of room, 20% the children's room, and 11% the living room. For parents, such a room where they can be without disturbance was their bedroom (94% of them mentioned it as being this type of room), and for children, the children's room was mentioned by 94%.

Room Restricted from Others

In this study, 62% of the participants said there is no part of their home where they do not want other family members to enter unless permitted, and 32% said there are such areas inside their home. As a restricted room, the parents' bedroom was most often mentioned (52%), followed by children's room (21%). Four out of 52 persons who answered this question mentioned that they would not like the family members to look into their cupboards or shelves. In addition, differences were indicated in the answers of parents and children: parents consider such areas mostly as their bedrooms (74%), and the children, their own rooms (100%).

The Overlap between the Rooms

This study also investigated the overlap between the separated, private, personalized space, and space restricted from others. We also tried to register how much these spaces overlap with the parts of the home where the family members try to be alone or where they sleep. One of the main conclusions is that the greatest overlap is between the rooms that are personalized, where one can be alone in privacy, and undisturbed. The least overlap was found with rooms closed from others, which is quite logical because only 1/3 of the respondents stated that there is this type of room in their home. As mentioned earlier, the parents considered their bedroom as a room that belongs to them that expresses their personality, where they do not want others to enter without permission. In addition, for wives a similar type a room is the kitchen and for husbands, their workroom or study.

Domestic Activities

Receiving Guests

Among the families who participated in this study, the guests were usually received in the largest room of the residence (answers are on the 5-point scale) ($x_{largest}$=3.66; X_{second}=1.42; X_{third}=1.74; X_{fourth}=1.67; X_{fifth}=1.80). In addition, the accompanying activities for receiving the guests are cleaning (X=3.37), cooking (X=3.25), and setting the table (X=3.77). The arranging of furniture is somewhat more common (X=1.65). 43% of the respondents also indicated that the place for receiving guests is dependent on which family member is receiving the guests. Receiving personal guests can take place in the second or third largest room in the apartment. This is especially apparent when it is the children's guests.

Eating

Relying on the data of the present study, eating takes place regularly in the kitchen; of the 85% who said that they ate at home, and the 85% of those who ate at home during the weekend declared that they did this in the kitchen. During holidays and birthdays, eating or dinners usually take place in the largest room of the apartment; 81% of the respondents who ate at home indicated so. We could also mention that breakfast and lunch during the weekdays is an independent activity; whereas, the evening meal (dinner or supper) is usually group activity. The highest scores for eating together where birthdays and festive days ($X_{birthdays}$ = 3.98, and $X_{festives}$ = 3.93 respectively on a 5-point scale). In this case, eating was not an isolated activity, as only 1% of the respondents mentioned that nothing else happened during the time that they ate. Usually, conversation (33%), listening to the radio (26%), watching TV (20%), and reading newspapers (17%) were mentioned as concurrent activities. Activities that were mentioned that were not included in the questionnaire, which could take place simultaneously were studying, playing games, and playing computer games.

In addition, the fact that the other activities usually do not interfere with eating (X=1.11; SD=.34) on a 3-point scale (1 = *usually interferes*, 2 = *sometimes interferes*; 3 = *usually does not interfere*) supports the understanding that eating is not an activity that needs to be separated from other undertakings.

Watching TV

All the families participating had a TV in their household. In 89% of the cases, watching TV took place in the largest room. In three families, it took place in the second or third room by the size. In the families participating in the survey, watching TV was rather more individual than it was a shared activity. However, at the same time watching TV was not separated from other kind of activities; 2% of the respondents indicated that nothing else happens while watching TV, the remaining considered conversation (34%), eating (19%), studying and reading (18%), and playing (14%) while watching TV rather normal activities. In addition, sleeping, ironing, listening to music, etc. were mentioned and not considered an interference with TV watching.

Working, Studying and Playing

Adults working at home was found to be quite common among the families studied (on a 4-point scale from 'never' to 'often' the mean was 2.66 (SD=0.92)). Most often professional work was done in the largest room of the apartment (47%), but could be done in all other rooms as well. The difference could be related to the number of rooms in the apartment; in 1-3-room apartment, about 58% of the respondents used the largest room for professional work and in apartments with four or more rooms, only 30% of the respondents did so. It is noticeable, that 70% of the women did their professional work in the kitchen (the average was 13%).

Professional work was somewhat separated from other activities at home; 31% of the respondents declared that they do nothing else simultaneously. Activities that could take place simultaneously were watching TV (27%), talking to each other (16%), and listening to the radio (14%). Respondents indicated that these activities did not interfere with their work activities (X=1.51 on the 3-point scale, SD = 0.57).

91% of the children participating in the study studied at home. Usually it took place in the second or third largest room of the apartment (80% of the cases). The other activities, which were reported as occurring in parallel, were the study activities of other children (18%), watching TV (18%), and listening to the radio (15%); at the same time about 32% of the respondents reported that nothing else happens in the same room during that time.

Children's play usually took place in the smaller rooms of the apartment or house (in 66% of the cases), but also in the largest room.

Sleeping

45% of the respondents did not have a special room intended only for sleeping. There were about 26% of the family members, primarily parents, who declared that nothing more takes place in their bedroom. Sleeping places were located primarily in the second (44%), third (27%), and fourth (13%) largest rooms in the apartments. Inclusive of all families, 12% of the respondents slept in the largest room, 48% of them lived in a one-room apartment, and 33% of them had two or more children living in a 3-room apartment. Of those who slept in the largest room, 86% were parents.

23% of the respondents declared that nothing else happens in the same room during sleeping hours. However, 22% mentioned watching TV, conversations (21%), and professional work (14%). Handicraft, reading, and listening to music were also mentioned. Usually, these activities were not disturbing for those people sleeping X=1.51 on a 3-point scale; SD=.58).

Being Alone

Taking into account the answers of all respondents, it appears that most of the respondents had the possibility to be alone at home when they desired it ($X=2.34$; $SD=.80$ on a 3-point scale). Nevertheless, we can bring out the difference between the possibilities of being alone in the families who lived in crowded situations (more than one person per room) and families who lived in non-crowded situations ($X_{high} = 1.98$; $X_{low} = 2{,}54$; $p<.01$). There were no statistically significant differences between men and women. In addition, no significant differences could be found between the evaluations of parents and children about their wishes to be at home alone even though the parents' ratings were somewhat lower.

The place at home where family members go when seeking solitude is most frequently the bedroom (58%). In addition to this, women, of whom 54% prefer the bedroom, also seek solitude in the kitchen (18%) and the bathroom (10%). Males, in addition to the bedroom (50%), prefer the study room (14%) and living room (12%). For children, the children's room seems to be the only refuge.

The Spatial Relations of the Activities

In addition to the descriptive statistics, the hierarchical cluster analysis was used in the current study in order to identify the similarities of activities, based on their location inside the home. In the analysis, only the activities situated in the rooms were taken into account. The activities taking place in other rooms (e.g. kitchen or bathroom) were not included in the analysis. Activities such as eating on weekends and eating on working days were left out from this analysis as the frequencies showed these activities to be located in kitchen.

The results of the cluster analysis are shown on Figure 13.2. It is possible to differentiate 3 (or 4) groups of activities. The first cluster includes following activities: *eating on birthdays*, *eating on holidays*, *receiving common guests* and the *guests of the wives*, and *watching TV*. The activities, which describe this cluster the best, include *eating on birthdays* and *eating on holidays*.

The second cluster consists of children's activities: *studying* in the case of the first and the second child, *sleeping* and *being alone* of the first child. The activities most similar to each other in this cluster are *the studying of the first child* and *studying of the second child*.

The third cluster consists of two sub-clusters: one of the subgroups includes activities like *working of both parents*, *being alone* of the husband and *playing of the children* (this cluster is best described by *husband's activities of working and being alone*); the second subgroup consists of activities like *sleeping of parents*, *wife being alone*, *second child being alone*, and *welcoming the guests of husband* and *children* (the most similar activities in this group are the sleeping activities by parents).

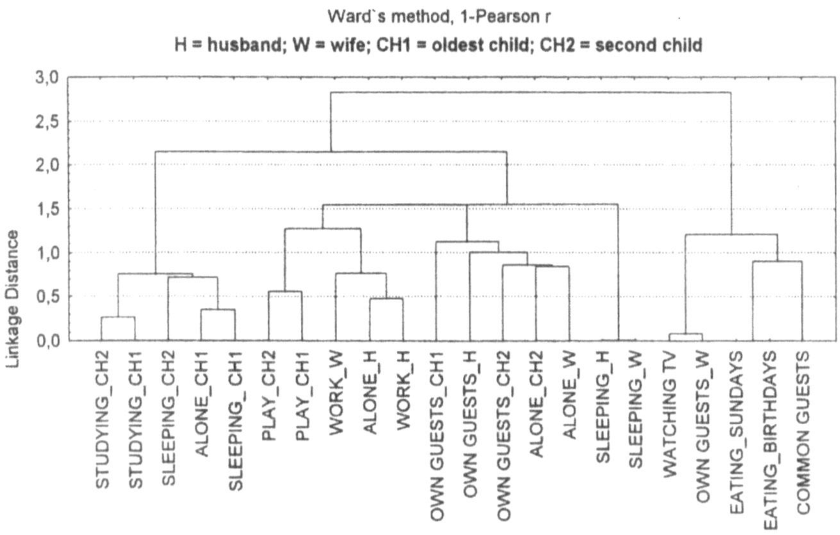

Figure 13.2 Tree diagram for 22 activities variables

The results of the cluster analysis presented here are in accordance with the results given by the frequency analysis. In addition, the cluster analysis identifies the difference between the spatial location of private activities (being alone, sleeping, working, and studying) and public activities (eating, watching TV, and receiving common guests). Private activities are located in the bedrooms of family members and public activities are located in the living room. In addition to that, the cluster analysis allows the differentiation between private activities. At first, it is possible to differentiate the activities of children from the activities of the parents; the cluster of children's studying and sleeping is different from the cluster of parents' working and sleeping. In the parents' cluster, it is also possible to differentiate working from sleeping. Therefore the cluster of parents is more diverse according to the activities that are located in it. This finding is to be expected, as it was shown earlier in the present study that all these activities can take place in all of the rooms at home, starting from the second largest room. For the reason mentioned above, it is expected that the clusters of children's and parents' activities are similar.

Discussion

The results of present study indicate that in taking into account the location of different activities at home and the attitudes of family members towards different rooms, it is possible to understand the home as a territorial system consisting of units that have different functions. Based on the results of the current study, it is possible to differentiate between at least two territorial units:

Public territory at home (living-room or the largest room), where the public activities are situated (receiving common guests, eating, watching TV) and were rarely seen as belonging to some family member, personalized and/or restricted from others.

Private territories at home (the bedrooms of family members, the kitchen in the case of the wife, and the study in the case of the husband), where the private activities were located (being alone, sleeping, receiving personal guests, working, studying) and were often seen as belonging to some family member, personalized, and restricted from others.

Other researchers (Sebba and Churchman, 1983; Omata, 1992, 1995; Heidmets, 1989; Giuliani, Rullo, and Bove, 1990; Altman and Chemers, 1980; Kim, 1997; Kimura and Sime, 1986) have also pointed out similar differentiation between public and private (sometimes called social/private) territories inside the home. Research has found that the bedroom of the family member is the place at home that is considered as belonging to oneself and is seen as the place reflecting one's personality; here the family members prefer to welcome their own personal guests and is where they usually go in order to be alone and avoid intrusion. The importance of the kitchen in the territorial behavior of the women has been pointed out before by Sebba and Churchman (1983). Heidmets (1989) has stated, based on his research, that in bigger homes the extra private room is usually given to one of the parents, most often to the husband, in order to use it as a study. The results of the present study are in accordance with this statement: they show that the extra study room found in bigger homes is usually considered the husband's study, personalized, and secluded from others.

Some of the studies have pointed out that in addition to private and public territory inside the home, the family territory, where the common activities of the family take place, should be differentiated as well (Omata, 1992, 1995; Giuliani, 1987; Altman and Chemers, 1980). Omata (Ibid.) sees this territory as the most important place in the Japanese home, a statement also supported by the fact that Japanese women do not consider their bedroom as personalized and controlled, rather the family room. Based on the results of the present study, the family room cannot be differentiated from the guests' room spatially. However, even though there appears to be spatial overlap of the guests' territory and public territory, these territories at home do not overlap temporally. Receiving guests and common activities of the family do not take place in the living-room at the same time, but receiving the guests requires many preparations (cleaning up, preparation of the food, etc), during which the private life of the family is put away from the eyes of the stranger for a while and the stage (in this case the family and the home) is set for the reception of the visitors.

In addition to the territorial units described above, private, public, and guests' territories, the differentiation between the territories of parents and children can be made in this study as well. In the case of parents, it is possible to see the spatial difference between working and sleeping territories. The finding that the territory for working will be separated from other territories at home if possible, is supported by previous research as well (Heidmets, 1989; Giuliani, 1987; Giuliani, Bove, and Rullo, 1990; Altman and Chemers, 1980).

Principally, it is also possible to differentiate the female and male territories inside the home on the basis of this study as well. Taking into account the place where different activities are located (receiving personal guests, being alone) and the attitudes of family members (belonging to myself, personalized, or room closed to others) it is evident that in addition to the bedroom, as most often mentioned in this relation, the kitchen can been seen as a female territory and the study as a male territory. In addition, other family members acknowledged these rooms as belonging to the male and female respectively. Therefore, in the current study, what is probably one of the most fundamental differences in the history of mankind is evident: the male is primarily connected to professional relations (working outside the home) and the female's field of activities is related to the household, even if she is also active outside the home (Heidmets, 1988).

The results of present study can also be interpreted in light of the conceptual framework presented in the theoretical part of the present paper, which states that the home consists of functionally different territorial units that all together regulate the openness/closedness of the family as a whole and the individuals living there. The understanding of how the different parts of a home function according to this regulation process is given Figure 13.3.

The home regulates the openness/closedness on three different levels: family *vs*. strangers, individual *vs*. family and individual *vs*. strangers. In the first case, the home as a whole has the function of separating the family from the outside world, where the role of uniting the family to the outside world is facilitated by the living-room. The living-room is the place inside the home where common guests of the family are received, which seems to be the clearest activity stating the family's willingness to open themselves up to the outside world. As many authors have pointed out, the living-room is also the place that is most actively decorated inside the home. It is the living room that can 'tell' the stories of the family to the strangers and at the same time, may also reflect the values and traditions that the family has accepted and that tie them to the community. Therefore, the living-room reflects the 'personality' of the family, not of the individuals who belong to the family even though according to the results of the current study, many people (mostly women) see the living-room as 'their' personalized room. This last mentioned finding can still be explained by the fact that most often the parents are the ones responsible for decorating the living-room, and as Sebba and Churchman (1983) have pointed out, people tend to see the rooms they are responsible for maintaining and decorating as the rooms reflecting their personality. From the other perspective, the parents are the ones inside the family who mostly represent the values of the family as a whole, and therefore, the reflection of their personality can be seen as a reflection of the personality of the family as well.

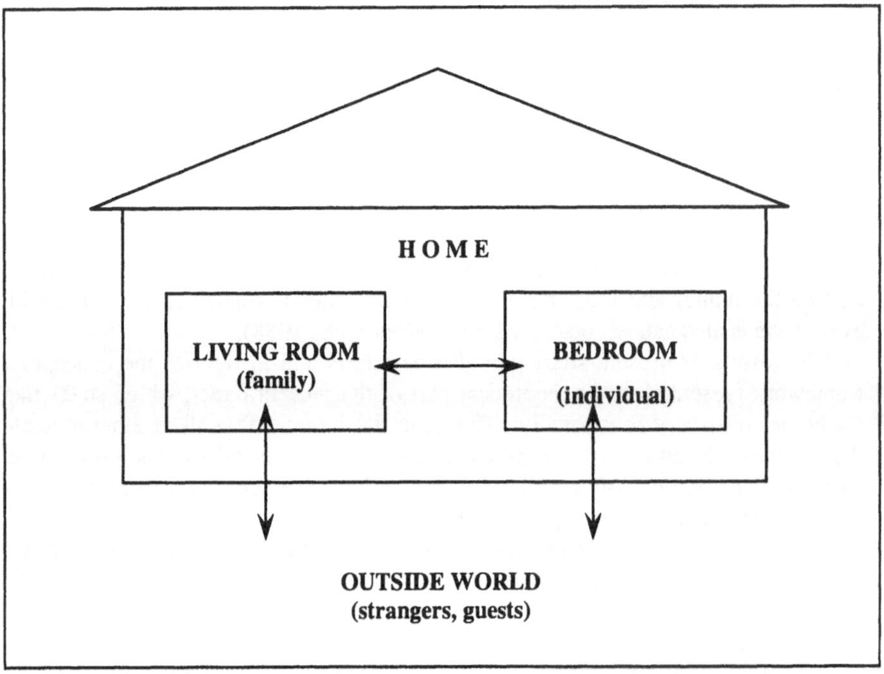

Figure 13.3 The home as a regulator of openness/closedness of the family and its individual members

The openness/closedness of the individual and family is regulated through the living-room and bedroom. The bedroom is a place at home in which the private activities (sleeping, being alone, working) take place, whereas the common activities of the family (watching TV, eating) take place in the living-room. Therefore, the bedroom can be seen as a territory offering family members the possibility to be separated from the rest of the family, whereas the living-room has the role of unifying the family. The fact that the living-room has a role of uniting the family, as well as a role of uniting the family to the outside world, is not contradictory because the living-room has these different functions at different times and in addition to that, receiving family's common guests can be seen as an activity that brings the family together.

In addition to the regulation of openness/closedness on the family *vs.* individual level, the bedroom regulates the openness/closedness also at the level of individual *vs.* strangers. This is the place seen as belonging to one's self: a place that is the reflection of one's personality and considered as the extension of one's self. The bedroom is also the room where specific individuals can welcome their own personal guests. The living-room, which reflects the family as a whole, seems not to be enough in this case. Opening up the bedroom to someone from the outside world, lets this guest come further into the personal and private world of an individual. Therefore, the bedroom has the same role on the level of the individual as the whole home has on the level of the family: the bedroom can be seen as a small copy of the home on the level of an individual (Heidmets, 1988). This kind of multiplication can be seen, for instance, in the case of receiving personal guests who are received both in the living room and in the bedroom.

Therefore, the results of present study can be interpreted in light of the regulation of openness/closedness – inside the home there are some territories with the function of separation and some territories with the role of unifying. Based on this study, the view is that it is mostly the bedroom and the living-room that have these roles, but obviously, this is not the only possibility. The last comment is also evident in the present study, which shows that the private room of the person can also be the kitchen or the study.

References

Altman, I. (1976). Privacy: A conceptual analysis. *Environment and Behavior*. 8(1), 7-29.

Altman, I., and Chemers, M. M. (1980). *Culture and environment*. New York: Cambridge University Press.

Brown, B. B. (1987). Territoriality. In D. Stokols and I. Altman (eds.), *Handbook of environmental psychology* .(pp. 505-531). New York: Wiley.

Canter, D. (1977). *The psychology of place*. New York: St. Martin's Press.

Edney, J. J. (1976). Human territoriality. In. H. M. Proshansky, W. H. Ittelson and L. G. Rivlin (eds.). *Environmental psychology: People and their physical settings* (2nd ed., 189-205). New York: Holt, Rinehart and Winston.

Giuliani, M. V. (1987). Naming the rooms: Implications of a change in the home model. *Environment and Behavior*. 19(2), 180-203.

Giuliani, M. V. (1988). The home as a territorial system. Paper presented at EUROPAN conference in West Berlin.

Giuliani, M. V., Rullo, G., and Bove, G. (1990). Socializing and privacy spaces inside homes: An empirical study. In H. Pamir, V. Imamoglu and N. Teymur (eds.), *Culture, space, history*. (pp. 129-138). Ankara: M.E.T.U., Faculty of Architecture Press.

Heidmets, M. (1988/1994). The phenomenon of personalization of the environment: A theoretical analysis. *Journal of Russian and East European Psychology*. 32(3), 41-85. (Originally published as: Феномен персонализации среды: Теоретический анализ. В кн. Х. Миккин (Ред.), *Средовые условия групповой деятельности* (с. 7-58). Таллинн: ТПедИ, 1988).

Heidmets, M. (1989). Персонализация в жилой среде: Эмпирическое исследование. В кн. Т. Нийт, М. Раудсепп and М. Хейдметс (Ред.), *Средовые условия развития социальных общностей* (с. 27-65). Таллинн: ТПедИ. [Personalization in the residential environment: An empirical study. In T. Niit, M. Raudsepp and M. Heidmets (eds.). *Environmental conditions for community development*. (pp. 27-65). Tallinn: Tallinn Pedagogical Institute (in Russian)].

Kim, M. H. (1997). The domestic space usage and behavior pattern in Belgium: With special reference to Dutch speaking area. In M. S. Amiel and J. Vischer (eds.) *Space design and management for place making* . (p. 172). Edmonton, OK: EDRA (Abstract).

Kimura, M., and Sime, J. D. (1986). Spatial identity boundaries in Japanese and English family homes. In J. W. Carswell and D. G. Saile (eds.). *Purposes in built form and culture research* (pp. 46-50). Lawrence, KS: University of Kansas.

Märtsin, M. (1999). *Ruumikasutus kodus* [The use of space at home]. Seminar paper. Tallinn: Department of Psychology, Tallinn Pedagogical University (in Estonian).

Niit, T. (1988). A methodological framework for studying families in dwelling environments. In H. van Hoogdalem, N. L. Prak, T. J. M. van der Voordt and H. B. R. van Wegen (eds.), *Looking back to the future*. (pp. 382-391). Delft: Delft University Press (.pdf full text version available from http://iaps.scix.net/cgi-bin/works/Home).

Omata, K. (1992). Spatial organization of activities of Japanese families. *Journal of Environmental Psychology*. 12, 259-267.

Omata, K. (1995). Territoriality in the house and its relationship to the use of rooms and the psychological well-being of Japanese married women. *Journal of Environmental Psychology*. 15, 147-154.

Rullo, G. (1992a). When the young remain in the parental home: Use of dwelling space and self-evaluation of living arrangements. In A. Mazis and C. Karaletsou (eds.). *Socio-environmental metamorphoses: Builtscape, landscape, ethnoscape, euroscape*. (pp. 94-98). Thessaloniki: Aristotle University of Thessaloniki.

Rullo, G. (1992b). Experience of the home among young adults in different living arrangements: Territorial patterns and satisfaction. In M. V. Giuliani (ed.), *Home: Social, temporal and spatial aspects*. (pp. 39-51). Milano: PFEd-CNR.

Sebba, R. and Churchman, A. (1983). Territories and territoriality in the home. *Environment and Behavior*. 15(2), 191-210.

Index